# BestMasters

Springer awards „BestMasters" to the best master's theses which have been completed at renowned universities in Germany, Austria, and Switzerland.

The studies received highest marks and were recommended for publication by supervisors. They address current issues from various fields of research in natural sciences, psychology, technology, and economics.

The series addresses practitioners as well as scientists and, in particular, offers guidance for early stage researchers.

Tamara Bernadette Aigner

# Photopolymerizable Porous Polyorgano-phosphazenes

## Degradable Matrices for Tissue Engineering

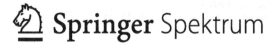 Springer Spektrum

Tamara Bernadette Aigner
Bayreuth, Germany

BestMasters
ISBN 978-3-658-09319-8          ISBN 978-3-658-09320-4 (eBook)
DOI 10.1007/978-3-658-09320-4

Library of Congress Control Number: 2015933839

Springer Spektrum

Printed on acid-free paper

Springer Spektrum is a brand of Springer Fachmedien Wiesbaden
Springer Fachmedien Wiesbaden is part of Springer Science+Business Media
(www.springer.com)

# Institutsprofil

Das Institut für Chemie der Polymere (ICP) an der Johannes Kepler Universität (JKU) Linz, das ich von 2007 an neu aufbauen durfte, fokussiert seine Forschung auf die Entwicklung funktionaler Polymere für den Einsatz in Drug Delivery und Controlled Release, im Bereich Tissue Engineering, in Hautpflegemitteln und Waschmitteln, als Träger von Katalysatoren sowie als Festphasen in Chromatographie und Extraktion. Vor allem werden für einige dieser Zwecke bioabbaubare Polyphosphazene maßgeschneidert. Bionik und Biomimetik spielen eine Rolle bei einem Projekt, bei dem es um neuartige DNA-Analoga geht, also Polymere, die später einmal in der Antisense-Therapie eingesetzt werden könnten. Polymere aus nachwachsenden Rohstoffen kommen auch zum Einsatz. Außerdem arbeitet das ICP an einigen Projekten, bei denen es um industrielle Aspekte von Polymeren geht. So sollen z.B. Kunststoffe für zukünftige thermische Solarkollektoren über chemische Kopplungen von Stabilisatoren an die Polymerketten derart modifiziert werden, dass die Kunststoff-Kollektoren deutlich längere Haltbarkeiten aufweisen.

Das ICP befindet sich mittlerweile in einer besonders fruchtbaren Umgebung, mit einem Polymerforschungsschwerpunkt an der JKU, mit neu hinzugekommenen Kunststofftechnik-Instituten, mit einer starken Kunststoff-Industrie in nächster Nähe und mit einer neu errichteten Medizinfakultät im Jahr 2014.

Univ.-Prof. Dr. Oliver Brüggemann
Institutsvorstand
Institut für Chemie der Polymere (ICP)

# Acknowledgement

This master thesis would not have come off without the contribution of many people. First I would like to thank my supervisor Univ.-Prof. Dr. Oliver Brüggemann for giving me the chance to work on this great project. Furthermore I thank my "Tissue Engineering Team": Dr. Ian Teasdale for his excellent guidance and his irrepressible optimism no matter what happened, Aitziber Iturmendi for her enthusiasm and spreading good cheer throughout the laboratory, Sandra Wilfert, PhD for her valuable ideas and lending me her ear everytime and Dipl.Ing. Martina Prambauer for introducing me so nicely to the laboratory. I also want to thank the other members of the institute for their help and the great time.

Moreover I thank our project partners without them the project would not have been possible. My special thanks go to Dr. Florian Hildner and Maria Rigau from Red Cross and Gbenga Olawale, MSc from BioMed zet for their great biological contribution funded by Regio 13 and to Transfercenter für Kunststofftechnik for their financial support.

Thanks to Dr. Wolfgang Schöffberger for the solid state measurement. The NMR experiments were performed at the Upper Austrian - South Bohemian Research Infrastructure Center in Linz, co-financed by the European Union in the context of the project "RERI-uasb", EFRE RU2-EU-124/100-2010 (ETC Austria-Czech Republic 2007-2013, project M00146).

I would like to thank my parents, my brother and my friends for their constant support and understanding. Last but not least I thank Thomas for being at my side all the time.

Tamara Bernadette Aigner

# Abstract

Tissue engineering is an emerging field combining biology, chemistry and materials science to repair, restore or regenerate living tissues. It is based on a matrix construct, commonly from a biodegradable polymer, which must fulfill the following requirements: mechanical stability to carry the cell load, interconnected porous structures to enable cell infiltration and communication, cell recognition sites to allow positive cell interaction, non-toxic degradation products and an adapted degradation rate so that cells can rebuild the extracellular matrix.

In this thesis we aimed to develop photopolymerizable matrices based on biodegradable poly(organophosphazenes). A crucial element of all these polymers was an allyl group to enable thiol-ene photochemistry. Thereby thiol-bearing molecules were reacted with the polymers to functionalize the poly(organophosphazenes) with several functional moieties to tailor the chemical, physical and biological properties. Moreover thiol-ene photochemistry was applied to crosslink the polymers to obtain mechanically strengthened networks. Blending agents were added to the crosslinking media, which influence the performances of the final materials in terms of mechanical stability, degradation rate and biological applicability. An interconnected porous structure was introduced using a novel photocrosslinking particulate-leaching technique.

The materials were characterized thoroughly using NMR and FTIR spectroscopy, GPC, CT, SEM and elemental analysis. Degradability is a main issue in tissue engineering, thus several degradation studies were performed and the structure of the polymers was adjusted to tailor degradability. This finally led to an amino acid containing side chain to fit biological systems. In order to assess their suitability as matrices for tissue engineering, cytotoxicity, adhesion and proliferation tests were performed. These biological tests showed promising results, paving the way for poly(organophosphazenes) in tissue engineering.

# Contents

# List of Figures

# List of Tables

# Abbreviations

| | |
|---|---|
| **ASC** | adipose-derived stem cells |
| **ATR-FTIR** | attenuated total reflectance fourier transform infrared |
| **bFGF** | basic fibroblast growth factor |
| **CT** | X-ray computed tomography |
| **DCM** | dichloromethane |
| **DMPA** | 2,2-dimethoxy-2-phenylacetophenone |
| **EC** | endothelial cells |
| **ECM** | extracellular matrix |
| **FITC** | fluorescein isocyanate |
| **GPC** | gel permeation chromatography |
| **HIPE** | high internal phase emulsion |
| **LDH** | lactate dehydrogenase |
| **PBS** | phosphate buffered saline |
| **PCL** | polycaprolactone |
| **PDGF** | platelet-derived growth factor |
| **PEG** | polyethylene glycol |
| **PGA** | polyglycolic acid |
| **PHB** | polyhydroxybutyrate |
| **PLA** | polylactic acid |
| **PLGA** | poly(lactic-*co*-glycolic acid) |
| **PPZ** | poly(organophosphazene) |
| **REDV** | arginine-glutamic acid-aspartic acid-valine |
| **RGD** | arginine-glycine-aspartic acid |
| **SEM** | scanning electron microscopy |
| **SFF** | solid freeform fabrication |
| **TFA** | trifluoroacetic acid |
| **TGF-$\beta$** | transforming growth factor beta |
| **THF** | tetrahydrofurane |
| **trithiol** | trimethylolpropane tris(3-mercaptopropionate) |
| **VEGF** | vascular endothelial growth factor |
| **YIGSR** | tyrosine-isoleucine-glycine-serine-arginine |

# 1. Aim of project

The aim of the project was to design a porous poly(organophosphazene) matrix suitable for tissue engineering. The increasing number of donor organs needed cannot be provided for anymore in this highly-developed era. Many patients are waiting for an allogenic transplant, thus for a young and healthy person to die. Even if the patient then receives the transplant, he/she has to take several drugs to suppress the immune system to avoid rejection of the organ. Additionally transplanted organs usually work properly for approximately 15 years only (depending on transplanted organ, age and general health situation).

Therefore tissue engineering is a future-oriented field combining skills from materials chemistry, biology and medicine. Several tissue types have already been studied like for example skin, blood vessels, nerves, bones, etc., but organs built from patient's cells on a scaffold are still a distant milestone. Several challenges need to be overcome first and this project was conducted to get a step closer to this goal.

In this project chemists, biologists and materials scientists worked together using poly-(organophosphazenes) as a basis for the scaffolds. Currently used natural materials like collagene or fibrinogen suffer from drawbacks like limited mechanical strength and irreproducible matrices. Synthetic materials like the frequently used polyesters or polycaprolactone show, amongst other drawbacks, problematic degradation rates, acidic degradation products or insufficient biocompatibility. The poly(organophosphazene) backbone is known to degrade under physiological conditions releasing neutral degradation products. By the addition of proper side chains the properties and therewith the degradation rate of the polymer can be tailored to be attractive to cells. Crosslinking can be used additionally to obtain mechanical strength.

The task of the chemistry group was to develop a novel poly(organophosphazene), which fulfills all indispensable requirements. The polymer must be biocompatible and show a positive cell interaction. The cells should adhere to the scaffold, differentiate on it and invade it. To allow cell invasion and communication the scaffold has to show an interconnected porous structure. This porous structure on the one hand requires a certain mechanical stability to carry the cells without collapsing. On the other hand it has to degrade at a suitable rate, so that cells can build up their extracellular matrix, and release only non-toxic and pH neutral degradation products. Additionally, the matrices should be synthesized and purified in large amounts.

1

# 2. Introduction

## 2.1 Tissue engineering

Tissue engineering is an interdisciplinary field applying "biological, chemical and engineering principles toward the repair, restoration, or regeneration of living tissues using biomaterials, cells, and factors, alone or in combination" [1].

### 2.1.1 Basic principle

There exist three main approaches to tissue engineering: 1. use isolated cells or cell substitutes to replace cellular parts, 2. use acellular biomaterials to mediate tissue regeneration and 3. use a combination of 1. and 2. usually by seeding cells on a scaffold *in vitro* producing a transplantable organ. [2, 3] The present project treats the third approach and the basic concept is shown in Figure 2.1. A polymer, in this case a poly(organophosphazene), is produced, which is biocompatible and degrades at a certain rate into non-toxic, pH-neutral molecules. This polymer may be functionalized to obtain cell recognition sites, thus allowing positive cell interaction. Then the polymer is crosslinked and a interconnected, porous structure is generated by the help of a porogen. The interconnected pores allow a high mass transfer and waste removal. At the same time cells from the patient are isolated and cultivated. These cells are then seeded onto the porous scaffold, providing mechanical support and determining the shape of the material. This "pre-tissue" is further cultivated in a bioreactor, where the cells proliferate, migrate and differentiate into the specific tissue. Once this tissue is engineered, it can be transplanted into the patient. The polymer scaffold degrades slowly, while it is replaced by extracellular matrix (ECM) components produced by the cells. [4, 5]

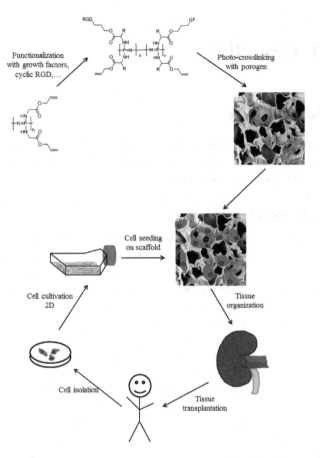

**Figure 2.1** – Basic principle: A polymer is functionalized and crosslinked with a porogen. Cells from the patient are cultivated and then seeded on the porous scaffold to obtain a functional tissue, which can be transplanted.

The reason for the considerable effort in the field of tissue engineering is obvious. Millions of patients are waiting for allogenic organ transplantations, which are limited by the number of donor organs available. All of them are on waiting lists and some eventually die before receiving an organ. [2] Also allogenic transplants bear the risk of disease transmission and are rejected by the immune system [6]. Thus several drugs must be taken to suppress the patient's immune system. Thereby, the general immune response is down-regulated, leading to an increase in infection risk. In comparison to that an engineered organ would resemble an autotransplantation, so the organ would be immunological tolerant and thus not be rejected [2].

Another problem with allogenic organs is the limited lifetime of approximately 15 years. In comparison to that, an artificial autogenic organ would lead to a permanent solution. Hence supplementary therapies would not be required, enabling a cost-effective treatment. [7]

The main challenges nowadays faced in tissue engineering are to find appropriate cell sources, design biocompatible materials and drug delivery systems and induce vascularization [6] to provide the cells with oxygen and nutrients and allow removal of waste products [5]. Especially the latter one is a reason, why up to now only engineered skin and cartilage are used in humans [8]. Skin is very thin so that no vasculatization is required and cartilage cells need only very little oxygen, therefore thick cross-sections (>1 mm) could be fabricated [5, 9, 10]. Although several tissues have entered clinical trials (bladder, blood vessels, bone, cornea, urinary structures and left mainstem bronchus [8, 11, 12]) and research is being carried out in many more (peripheral nerves, muscle, connective tissues like tendons and abdominal wall, joints, heart components, breasts, small intestine, esophagus, pancreas, liver and tracheal constructs [12, 13]), the vascularization problem is not yet solved and hence the search for an appropriate biomaterial and fabrication techniques goes on.

## 2.1.2 Polymer scaffolds

Isolated cells are able to reconstruct tissue structures only, when they obtain guidance by a template, because cells cannot grow in certain three-dimensional orientations by themselves, they rather form random 2D layers. Hence, to obtain an anatomically correct tissue, cells must be seeded and cultivated on porous matrices, usually referred to as scaffolds. [2, 5] In the body the ECM, surrounding cells and soluble factors perform this guidance role, giving the cells a controlled microenvironment [14], which plays a considerable role in cell fate and function [6]. Therefore, a scaffold used for tissue engineering should mimic the ECM, including biological factors like growth factors.

### From ECM to polymer scaffolds

Understanding the nanocomposite nature and working principle of the ECM is the first step to produce suitable biomaterials. The ECM is a very dynamic and hierarchically organized system, which regulates essential cellular functions like adhesion, migration, differentiation, proliferation and morphogenesis. [4, 15] It provides structural support, a stem cell niche, epithelial-mesenchymal interactions and modulates the availability of morphogenetic and growth factors. Structural support comes mainly from collagens, which are triple helical proteins giving compressive and tensile strength to animal tissue. It also possesses anchors for cell adhesion. [15] This collagen network is interwoven with elastin [4], which provides elasticity to allow the tissue to stretch and later return to its original state [16]. The collagen-elastin network is covered by adhesive proteins like laminin and fibronectin. These form the connection between the

5

anchors of collagen and the integrin receptors of the cells. [4] The integrin receptors initiate intracellular signalling cascades, whereby important cell behaviours like growth, shape, migration and differentiation of cells are regulated [17]. Proteoglycans, which consist of a core protein with a polysaccharide shell, are important for the ECM assembly and mediate cell adhesion and motility. These polysaccharide chains are classified as keratin sulfate, dermatan sulfate, chondroitin sulfat and heparan sulfate. [15] Also, other non-protein bound polysaccharides are present in the ECM like hyaluronic acid. In general, polysacchrides fill the interstitial space between the network fibers to assist in form of a compression buffer against stress and serve as growth factor depot. [4, 18] Different tissues have different requirements regarding ECM, so it is understandable that the ECM composition and organization varies between tissue types to ensure tissue specific morphology, shape, function. [4]

For the successful fabrication of scaffolds resembling the ECM, the following requirements should be fulfilled: 1. interconnected porous structure of adequate size to allow tissue vascularization and integration, 2. materials possessing controlled biodegradability or bioresorbability, so that the scaffold can be replaced by ECM, 3. appropriate surface chemistry to allow cellular adhesion, proliferation and differentiation, 4. appropriate mechanical properties fitting to the site of implantation, 5. no induction of any adverse response and 6. easy fabrication into various sizes and shapes. [5, 19]

Considering these requirements, several attempts using natural materials have been made. In general, compounds found in the animal ECM were chosen for instance collagen, fibrinogen, proteoglycans, hyaluronic acid and hydroxyapatite, but also other natural occurring substances found in plants or insects like starch, alginate, chitin, soy or silk have been considered [6, 20]. The significant advantage of these materials is the presence of recognition sites, which facilitate cell attachment and maintain the differentiated function of the cells. Unfortunately, natural materials suffer from several drawbacks for example poor mechanical properties, inability to tailor degradation rates and limited control over other physio-chemical properties. Further problems occur in the fabrication like batch-to-batch variations and hence scale-up difficulties, challenges in sterilization and purification and the possibility of induced immune responses, if different sources are used. [2, 6, 20–22]

Since natural materials show unacceptable disadvantages, synthetic polymers are gaining popularity. Mainly aliphatic polyesters are used as for example polylactic acid (PLA), polyglycolic acid (PGA), their copolymers poly(lactic-co-glycolic acid) (PLGA), polycaprolactone (PCL) or polyhydroxybutyrate (PHB). Polyesters show considerable advantages compared to natural materials as they possess good mechanical strength, can be processed with known methods, the degradation rate can be tailored to an optimum, the molecular weight can be controlled better than in natural polymers and they are biocompatible. Yet, synthetic materials show a poor inherent bioactivity as they lack cell-recognition sites. Besides, most have a rather hydrophobic and hence, unattractive surface for cells. [2, 5, 6, 20, 23] These disadvantages may be overcome by modifying polymers with critical amino acid sequences from natural materials, as explained in Chapter 2.5. The crucial disadvantages of these polyesters are the bulk degradation kinetics instead of surface erosion, so the polymers fall apart giving the cells no chance to slowly replace the ECM, and acidic degradation products, which then accelerate degradation and reduce the pH of the surrounding. [2, 5, 6, 14, 23, 24] The human body reacts very sensitive to pH-changes and already slight differences may cause severe troubles. Hence, a scaffold with acidic degradation products is not suitable for most applications.

It is a demanding task to effectively organize cells into a tissue, which morphologically and physiologically resemble those *in vivo*. Especially as not all signaling factors driving tissue assembly have yet been identified. Cells are guided by the interaction with different characteristics of the ECM, namely its topography [25], mechanical properties like stiffness, elasticity and viscosity [26–28] and the concentration gradients of ECM molecules [29] or immobilized growth factors [4, 30]. Hence, not a single biomaterial can fulfill all these requirements, but rather a "multi-component polymer system" [20] must be established.

### 2.1.3 Vascularization and signaling molecules

Tissue engineering still has several challenges. Alongside preparing a suitable scaffold and finding a renewable source of functional cells, introducing an artificial vascular system is a crucial point. [14] Cells survive only a few $100\,\mu$m away from a capillary [2] and hence also do not migrate more than $500\,\mu$m into an engineered scaffold [5]. Beyond this distance, the cells are not sufficiently supplied with oxygen and nutrients and waste products cannot be removed. [5, 6, 11] The problem hereby is that the exchange of gas, nutrients and waste occurs only at the surface, so with increasing tissue size the volume increases by $r^3$, whereas the surface increases only by $r^2$. [11] Hence, for larger tissues a branched vascular system is necessary to increase the overall surface and allow the vital exchanges.

Neovascularization, the new formation of blood vessels, can be induced in endothelial cells (ECs) by angiogenic factors. These factors, naturally present in the ECM, initiate angiogenesis by inducing endothelial cell proliferation and migration (basic fibroblast growth factor - bFGF and vascular endothelial growth factor - VEGF), recruiting pericytes and smooth muscle cells (platelet-derived growth factor - PDGF) and depositing ECM to stabilize the newly formed vessels (transforming growth factor-$\beta$ - TGF-$\beta$). [6, 31–33]

These angiogenic as well as other signaling molecules can be added to the culture medium or can be incorporated within the scaffold directly by covalent immobilization or by encapsulation. [6, 34, 35] Covalent bound growth factors, hormones, cytokines and other chemicals show prolonged signaling as their endocytosis is inhibited. Several covalent bound factors like VEGF, bFGF, ephrin-B2 and ephrin-A1 showed a prolonged and improved angiogenic performance. [6, 36–40] Encapsulated factors could be tuned to obtain a controlled spatiotemporal release, this would allow to design an ideally localized signal molecule composition at each time point. [8] Therefore, a successful polymer scaffold candidate should allow the addition of signaling molecules to enable successful tissue regeneration and growth.

## 2.2 Poly(organophosphazenes)

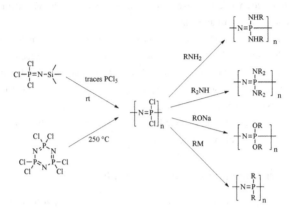

**Figure 2.2** – Cationically catalyzed living condensation polymerization and ring-opening polymerization give poly(dichlorophosphazene), which can subsequently be reacted with different nucleophiles.

Poly(organophosphazenes) are a type of "hybrid inorganic-organic polymers", hence macromolecules with an inorganic backbone and organic side chains. [41] The inorganic backbone consists of the repeating unit -P=N-, which is commonly arranged in linear chains. [42] The synthesis of the polymer is a two-step reaction sequence as shown in Figure 2.2. First the precursor poly(dichlorophosphazene) is prepared either by a ring-opening polymerization at $250\,^{\circ}$C or by a living cationic polymerization [43, 44], which was used in this project. Then the chlorine atoms on the phosphorus are replaced by organic nucleophiles. [43–45] The starting material of the ring-opening polymerization is the cyclic trimer, which is reacted to the linear polymer in molten state or in solution [43, 44]. The resulting polymer has a high molecular weight ($\sim$15,000 repeating units) and a broad molecular weight distribution [42]. In comparison to the ring-opening polymerization, the cationically catalyzed living condensation polymerization has a narrow molecular weight distribution and the chain length can be controlled in a certain range (10-200 repeating units) [46, 47] by varying $PCl_5$ to monomer ratio. [42, 48] Other rarely used low-molecular weight synthesis routes are condensation reactions between $PCl_5$ and $NH_3$ [49] or condensation polymerization of $Cl_3P=NP(O)Cl_2$ [50]. The chlorine atoms of the very reactive poly(dichlorophosphazene) are substituted by primary or secondary amines, metal alkoxides or aryloxides or organometallic reagents. Also different types of side groups may be added on the same molecule either by a sequential or simultaneous replacement reaction. [42, 51]

8

This flexibility and variety in possible side chains allows to tailor physical, chemical, mechanical and biological properties to suit the required task. Far more than 700 different poly(organophosphazene) have been fabricated ranging from low-$T_g$ elastomers to high-$T_g$ glasses, and from microcrystalline fibers and films to biocompatible materials. The polymers are used for optical and photonic developments, energy storage and generation for instance in lithium ion or proton conductive polymers, aerospace materials, micelles, functional surfaces, nanofibers and in biomedicine. [42,45,48,51,52] The latter application requires biocompatible and biodegradable poly(organophosphazenes) with hydrolytically labile groups like amino acid esters, glycolate or lactate esters, steroidal residues, imidazolyl, glycosyl or glyceryl groups. [48,52] Laurencin *et al.* [53] were the first group to seed cells on a poly(organophosphazene) for bone tissue engineering studies and suggested that poly(organophosphazenes) are suitable candidates to construct cell-polymeric matrices for tissue regeneration [54].

Since then, more and more poly(organophosphazenes) for biological applications have been examined. Thereby, the degradation is a major issue. It can follow a bulk or surface degradation mechanism or a combination of both. [52] The degradation products of poly(organophosphazenes) are comprised of the biologically friendly non-toxic natural buffer phosphate and ammonia and the side group. [52,55] Generally, the more hydrophilic a poly(organophosphazene) is, the faster it degrades and, on the other hand, the more hydrophobic a side chain is the more it protects from hydrolytical cleavage.

The side chain has a major impact on the properties, thus, it must be chosen carefully. The side group has, for example a tremendous impact on the hydrophilicity of the polymer. The hydrophilicity in turn tailors the degradation rate ranging from hours to years at 37 °C [52,54,56, 57] The side group also has a considerable impact on the mechanical properties as for instance bulky side groups increase backbone rigidity. [45] Diverse forms of crosslinking using the side chains are frequently performed to increase stiffness and thermal stability. [48] Also, "smart" poly(organophosphazenes) have been fabricated, which respond to certain external stimuli like pH and temperature changes as for instance polar functionalities, which show a pH dependent solubility and swelling behavior. [52,58] Thus, poly(organophosphazenes) offer a very versatile and complex system, which can be adapted for many different applications and it is a unique class of polymers with a vast potential for tissue engineering applications [52].

## 2.3  Thiol-ene photochemistry

Thiol-ene photochemistry works via a radical reaction mechanism as shown in Figure 2.3. The addition is facilitated by a photoinitiator and proceedes through a thiol radical, which attacks the double bond of an alkene, acrylate, maleimide or norborane to form the thiol-ene product [59]. Thiol-ene chemistry has several benefits as it shows tolerance to different reaction conditions, solvents, oxygen and even water, it has an explicitly defined pathway giving quan-

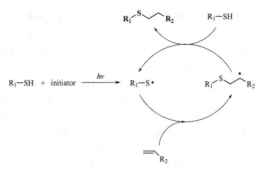

**Figure 2.3** – Mechanism of thiol-ene photochemistry

titative yields and it is regioselective. Furthermore, nonchromatographic purification methods can be used and the usage of toxic metal cations can be avoided. [59–62]

This photochemistry can be used to crosslink polymers like poly(organophosphazenes) to obtain the desired mechanical properties of the given polymers. [63] Thereby, a molecule possessing several thiol groups can be reacted with double bonds of the side chain of a linear poly(organophosphazenes) under UV-light. [64] Therewith, highly viscous poly(organophosphazenes) can be turned into firm polymer matrices without harming the -P=N- backbone as this is resistant to UV-irradiation [65].

Thiol-ene chemistry cannot only be used to crosslink poly(organophosphazenes), but also to introduce post-polymerization modifications. Especially unprotected multifunctional nucleophilic reagents can hardly be added selectively to poly(dichlorophosphazene) as this would lead to crosslinking [66–69]. Additionally, steric hindrance may be problematic, leaving unreacted chlorine atoms in the chain, which increase the degradation rate. Mixed macrosubstitution of the chlorine atoms would be a possible solution, but the considerable drawback hereby is that it is impossible to obtain repeatable results. Thus, it is reasonable to transfer the chlorine atom into an alkene, which can further be functionalized using interesting thiol bearing molecules. [66]

## 2.4   Pore generation - matrix formation

The polymer scaffold should guide cell growth and organization and allow the cells to invade and communicate. Furthermore, the diffusion of oxygen, nutrients and waste needs to be guaranteed to enable cell survival. [2, 6] Therefore, a 3D network with a highly interconnected porous structure is essential, whereby the pore size, pore number and pore connectivity are important parameters. If the pores are too large($\mu$m-mm), vascularization is impossible as endothelial cells cannot bridge pores larger than a cell diameter [70]. On the other hand, if the pores are too small (<100 nm) gas-exchange and nutrient supply is hindered. Thus, the porosity must be adapted to fit the integrity and mechanical properties of the material and the cellular effects on it. [6, 71–73] Another important advantage of porous scaffolds is the fact that these can be easier washed and thus cleared from residual solvents, reagents, etc.

Nowadays many different methods can be used to obtain a porous scaffold. In the following, a selection of several methods is presented shortly:

**Solid Freeform Fabrication:** Here, the 3D object is first created as a computer model and then built up layer by layer from bottom to top. [5] The following five techniques described are a type of Solid Freeform Fabrication (SFF).

**3D microprinting:** This technique works with a conventional ink jet printing technology and ejects a binder onto a polymer powder surface. This binder is able to dissolve and join adjacent polymer particles. After finishing the layer, fresh polymer powder is added and the process is repeated. Unbound powder acts as support during fabrication and can be removed afterwards. [74]

**Stereolithography:** Here, a liquid photocurable monomer is selectively polymerized by an UV-laser beam. After every layer, the elevator with the model is lowered so that the model is again covered with monomer. When the fabrication process is finished the model is lifted and support structures are removed. [75]

**Fused deposition modelling:** The head extrudes a fiber of polymeric material, which is deposited on an elevator. The model is created layer-by-layer by lowering the elevator until it is finished. [76]

**3D Plotter:** The extruder head can move in all three dimensions to deposit the liquid or paste like polymer, oligomer or monomer into a liquid medium. Solidification occurs when it comes in contact with a previous layer or a substrate. [77]

**Phase-change Jet Printing:** This printer has two ink-jet heads. One deposits the material for the actual model, whereas the other deposits a support. The model is build layer-by-layer in a drop-on-demand fashion. When the printing process is finished, the model is immersed in a solvent to dissolve the support leaving the scaffold. [78]

**Nanofiber electrospinning:** It is a simple and useful nanofabrication technique to get complex 3D matrices [8]. Here, interwoven fibers are created by controlling the deposition of polymer fibers on a substrate using an electric field. Thereby, a highly interconnected porous network is produced. Unfortunately, the fibers produced are at the upper limits of the fiber size in ECM (commonly between 50 and 500 nm). Also, the cells often cannot infiltrate the matrix, therefore hybrid scaffolds were made using a combination of electrospinning and 3D microprinting. [4]

**Molecular self-assembly of fibers:** This method can produce very thin fibres as the process of ECM assembly is mimicked [8]. This technique relies on the pH-induced organization of different components into a stable and ordered network with pre-programmed non-covalent bonds [79]. Very often a peptide amphiphile is used [79] to obtain a fiber diameter of about

10 nm and pore sizes ranging from 5 to 100 nm, so considerable smaller than the pores obtained by electrospinning [80]. [4]

**Gas foaming:** Here, the polymer is saturated with $CO_2$ at high pressure. Then, by bringing the $CO_2$ pressure back to atmospheric level, the solubility of $CO_2$ in the polymer is lowered resulting in nucleation and growth of gas bubbles giving rise to large pores (100-500 $\mu$m). Fiber-reinforced networks were produced by adding fiber polymers to the polymer matrix. [81]

**Phase separation:** A polymer is dissolved in molten phenol or naphthalene. Then the temperature is decreased to get a liquid-liquid phase separation and after quenching a two-phase solid is obtained. The solvent is removed leaving a porous scaffold. [82]

**Freeze drying:** The principle behind this method is that a polymer solution is frozen first to produce ice crystals, which are then removed by freeze-drying leaving a porous polymer scaffold. [83] The pore size depends on the pH and freezing rate, a fast freezing rate for instance gives smaller pores. [84]

**Emulsion freeze drying:** Here, ultrapure water is added to a polymer dissolved in an organic solvent immiscible with water, e.g. methylene chloride. The two phases are homogenised forming a water-in-oil emulsion. Then the emulsion is quenched in liquid $N_2$ and freeze-dried to obtain a porous scaffold with for instance a pore diameter of 13-35 $\mu$m using PLGA. [85]

**Sintering:** Polymer microspheres are heated above glass transition temperature. The beads become rubbery and are bound together. [86] Another approach is to use a solvent/non-solvent technique, where the surface of the particles is dissolved and is susceptible to bind to adjacent microspheres. [87]

**PolyHIPE:** To produce poly high internal phase emulsion (HIPE) materials the photoreactive monomers, a surfactant and a photoinitiator are mixed with an organic solvent giving the oil phase. Then, water is carefully added under stirring to produce a water-in-oil emulsion. This emulsion is photo-cured under UV-light. Then the water is removed leaving a highly interconnected porous material with a void diameter between 34-125 $\mu$m. [63]

**Melt moulding:** Here, a mould is filled with polymer powder and gelatine microspheres. The mould is heated above the glass-transition temperature of the polymer and is set under increased pressure so that the polymer particles are bond together. Then the gelatine is leached out in a water bath leaving the polymer scaffold. [88]

**Solvent-casting particulate-leaching:** A polymer is dissolved in an organic solvent and a salt is added and dispersed throughout the solution. The solvent is evaporated giving a matrix with embedded salt particles. This matrix is transferred to a water bath, where the salt leaches out giving a porous matrix. [89,90]

**Photocrosslinking particulate-leaching:** This technique was invented for this project and takes a cue from the solvent-casting particulate-leaching and the polyHIPE method. Hereby, a polymer and a trivalent crosslinking agent are dissolved in an organic solvent. Then salt is added as a porogen and the suspension is thiol-ene photocrosslinked under UV-light to gain mechanical strength. After UV-curing, the salt is removed in a water bath leaving a porous scaffold for use in tissue engineering.

## 2.5 Polymer functionalization

One major task of tissue engineering scaffolds is the stimulation of specific cell responses at the molecular level. Thereby, specific cell interactions are created directing cell fate, meaning cell attachment, proliferation, differentiation and the production and organization of ECM. [20, 91] The addition of certain functional motifs can even promote a rapid differentiation of certain cell types, while the differentiation of other cell lineages is discouraged. [92]

The attractiveness of scaffolds can be enhanced by coating them with a layer of ECM proteins like collagen, fibronectin or hyaluronic acid. These layers help to turn the matrices into hydrophilic scaffolds and provide anchors for cells. The problem hereby is that the interactions of cells with ECM are very complex and complicated. Thus, to avoid unwanted reactions often only short peptides derived from ECM proteins are used. These are the most primitive subunits, which are important for cell attachment and proliferation. [6] YIGSR (tyrosine-isoleucine-glycine-serine-arginine) is a laminin-derived peptide, which was for instance able to enhance endothelial cell (EC) adhesion and proliferation, while platelet adhesion was decreased [93]. A fibronectin-derived sequence is REDV (arginine-glutamic acid-aspartic acid-valine) binding integrins found on ECs, thereby assisting in specific adhesion and spreading [94]. By far the most common short peptide sequence is RGD (arginine-glycine-aspartic acid). RGD is present in many ECM proteins and enhances cell adhesion via binding of integrin receptors [6, 95, 96]. Most RGD-containing ECM proteins are helices and RGD is located at the tip of a loop in the cell-binding domain [97]. Thus, a cyclic form of RGD resembles natural conditions, which leads to an enhancement of cell adhesion by 240 times compared to the linear RGD. [98] Though, functionalization of scaffolds is not only about the addition of certain peptide sequences. It can also be used to render the physical and chemical properties of the scaffold like hydrophobicity/hydrophilicity or the degradation rate. [6]

Several methods may be used to immobilize functional groups, peptide sequences and signaling molecules like laser tethering [99] or microcontact printing [100] for RGD. Also, weaker electrostatic interactions can be used, for instance, by mimicking heparan sulphate, which then binds growth factors or cytokines [101]. In this project, thiol-ene photopolymerization was used to render the properties of the polymers by the addition of several functional groups like acids and esters to make the polymer hydrophilic, of fluorescein for visualization and of glutathione for biocompatibility.

# 3. Experimental

## 3.1 Materials and methods

Polymer synthesis and modification were carried out under inert atmosphere using standard Schlenk line techniques or under argon in a glovebox (MBRAUN). Phosphorus pentachloride was purified by sublimation and stored under argon. Triethylamine was distilled and stored over molecular sieves. Phosphonitrilic chloride trimer, divinyl adipate, ethylene glycol allyl ether, ethyl-3-mercaptopropionate and anhydrous toluene (toluene anh.) were bought from TCI Europe. Glutathione and all standard solvents were bought from VWR. Jeffamine-1000 was received from Huntsman. All other chemicals, anhydrous dichloromethane (DCM anh.) and tetrahydrofurane (THF anh.) were purchased from Sigma Aldrich and used without further purification. Phosphonitrilic chloride trimer with allylamine as a side chain was produced by Aitziber Iturmendi (lic. en Quimica) according to [102], herein after referred to as Trimer **1**. Phosphonitrilic chloride trimer with L-serine as a side chain reacted with pentenoic anhydride was also prepared by Aitziber Iturmendi according to Killops *et al.* [62], herein after referred to as Trimer **5**. Poly(organophosphazene) with ethylene glycol allyl ether and glycine ethyl ester was produced by Aitziber Iturmendi according to [102], herein after referred to as Polymer **8**.

A Rayonet Chamber Reactor equipped with a UV-lamp from Camag (254 nm) was used to carry out the photochemical reactions. The glass vials used for these photochemical reactions have a wavelength cutoff of $\sim$300 nm. Solution-state NMR spectroscopy was performed on a Bruker digital Avance III 300 MHz NMR-spectrometer using 300 MHz for $^1$H NMR, 81 MHz for $^{13}$C NMR and 75 MHz for $^{31}$P NMR spectroscopy. A Bruker DRX 500 was used for solid-state NMR analysis. FTIR measurements were conducted using a Perkin Elmer Spectrum 100 FTIR spectrometer equipped with an ATR accessory.

All biological tests were performed by our project partners namely DI Dr. Florian Hildner and Maria Rigau from Red Cross Blood Transfusion Service of Upper Austria, Ludwig Boltzmann Institute for Experimental and Clinical Traumatology, Austrian Cluster for Tissue Regeneration (Linz, Austria) and Gbenga Olawale, MSc from BioMed-zet Life Science GmbH (Linz, Austria). Red Cross used adipose-derived stem cells (ASC) and BioMed zet used human peritoneal mesothelial cells (HPMC). Gold sputtering of the polymer samples for scanning electron microscopy (SEM) was carried out by Markus Gillich, University of Applied Sciences Wels (Austria) and were subsequently imaged by Stefan Schwarzmayer, ARS Electronica Center (Linz, Austria). X-ray computed tomography (CT) was done by Dietmar Salaberger and Christian Hannesschläger, University of Applied Sciences Wels (Austria). Elemental analysis was performed by Johannes Theiner, University of Vienna, Mikroanalytisches Labor (Vienna, Austria).

## 3.2 Trimer synthesis

**Trimer 2**

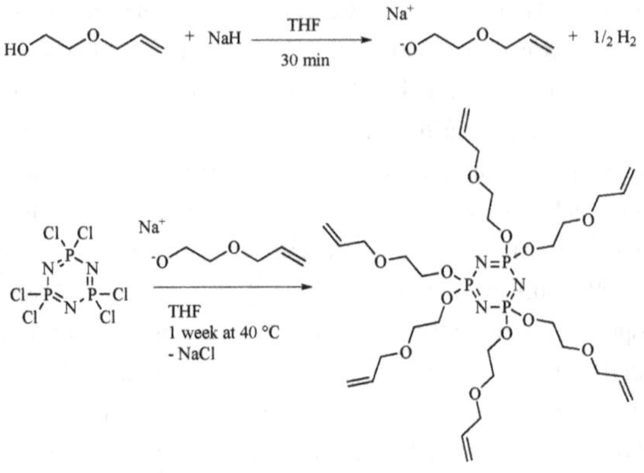

**Trimer 2**

**Figure 3.1** – Synthesis route for Trimer **2**

Ethylene glycol allyl ether (2 g, 5.8 mmol) was slowly added to a cooled suspension of NaH (1.40 g, 35 mmol) in 20 ml THF anh. under inert conditions. The reaction mixture was allowed to reach room temperature and stirred for 30 minutes. Phosphonitrilic chloride trimer (2 g, 5.8 mmol) dissolved in 60 ml THF anh. was added to the cooled mixture. The reaction was heated to 65 °C for 9 hours and then on at 40 °C overnight. After one week, the reaction was stopped and the suspension was centrifuged to remove NaCl. The solvent was removed under reduced pressure leading to a slightly pink gel. The reaction mechanism is shown in Figure 3.1. Yield: 4.1 g (96 %) $^1$H NMR (CDCl$_3$): $\delta$ = 3.63-3.66 (t, 2H), 4.00-4.10 (m, 4H), 5.15-5.29 (dd, 2H), 5.82-5.95 (m, 1H) ppm. $^{31}$P NMR (CDCl$_3$): $\delta$ = 18.01 ppm. FTIR (solid): $\nu_{max}$ = 3351 (N-H), 2925 (C-H), 1729 (C=O), 1178 (P=N) cm$^{-1}$.

**Trimer 3**

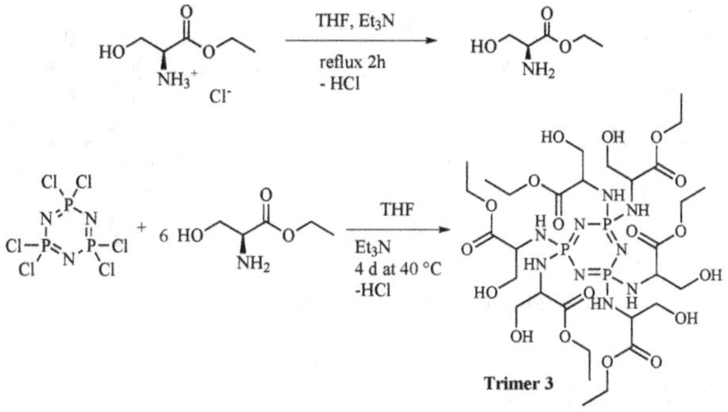

**Figure 3.2** – Synthesis route for Trimer **3**

According to Morozowich *et al.* [103], L-serine ester HCl (13.21 g, 77.7 mmol) was suspended in 180 ml THF anh. and 10 ml Et$_3$N and refluxed for 2 hours under inert conditions. The suspension was filtered and slowly added to the phosphonitrilic chloride trimer (3.02 g, 8.6 mmol) dissolved in 20 ml THF anh. and 10 ml Et$_3$N. The reaction was stirred 4 days at 40 °C. Then, the suspension was filtered and the solvents were removed under reduced pressure. The raw product was purified by dry column vacuum chromatography (heptane → ethyl acetate) yielding a yellow highly viscous oil. The reaction mechanism is shown in Figure 3.2. Yield: 0.4 g (5 %) $^1$H NMR (CDCl$_3$): $\delta$ = 1.27 (m, 3H), 3.92-4.51 (m, 5H) ppm. $^{31}$P NMR (CDCl$_3$): $\delta$ = 35.02 ppm. FTIR (solid): $\nu_{max}$ = 3263 (N-H), 2982 (C-H), 1733 (C=O), 1197 (P=N) cm$^{-1}$.

### 3.2.1 Trimer modification

**Trimer 4**

According to Rim and Son [104], Trimer 3 (0.20 g, 0.2 mmol) and K$_2$CO$_3$ (0.53 g, 3.8 mmol) were dissolved in 100 ml acetone under inert conditions. Carefully, acryloyl chloride (0.31 ml, 3.8 mmol) was added and the reaction mixture was refluxed overnight. The reaction was stopped and filtered. Then, the solvent was removed under reduced pressure. The crude product was purified by dry column vacuum chromatography (heptane → ethyl acetate → methanol) to obtain an orange oil. The reaction mechanism is shown in Figure 3.3. Yield: 33.4 mg (12.5 %) $^1$H NMR (CDCl$_3$): $\delta$ = 1.27 (m, 3H), 3.65-4.6 (m, 5H), 5.72-6.40 (m, 3H) ppm. FTIR (solid): $\nu_{max}$ = 3345 (N-H), 2984 (C-H), 1728 (C=O), 1641 (C=C) 1199 (P=N) cm$^{-1}$.

17

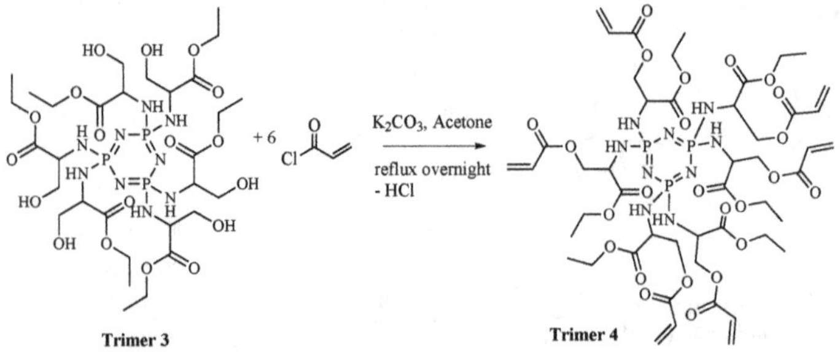

**Figure 3.3** – Synthesis route for Trimer **4**

## 3.3 Polymersynthesis

### 3.3.1 One-pot polymerization

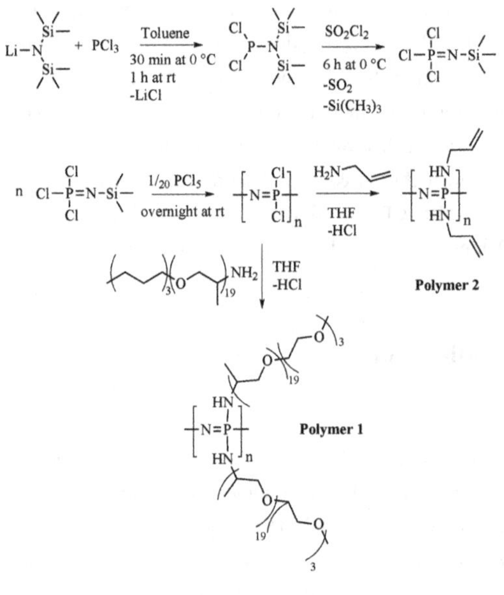

**Figure 3.4** – One-Pot polymerization route for Polymer **1** and **2**

**Polymer 1 and 2**

According to Wang [105], LiN(SiCH$_3$)$_2$ (5.06 g, 29.9 mmol) was dissolved in 110 ml toluene anh. and cooled to 0 °C. PCl$_3$ (2.61 ml, 29.9 mmol) was added carefully and then, the suspension was stirred for 30 min at 0 °C followed by 1 h at room temperature. The mixture was again cooled to 0 °C before $SO_2Cl_2$ (2.42 ml, 29.9 mmol) was added slowly. The reaction was allowed to proceed for 6 h at 0 °C. Then, $PCl_5$ (0.31 g, 1.5 mmol) dissolved in 30 ml toluene anh. was added and the suspension was stirred overnight at room temperature. The next day, the yellowish-turbid mixture was filtered through Celite, which was washed with toluene anh. The solvent was removed under reduced pressure. 50 mg of the yellow residue were weighed in a separate vial. Then, each part was redissolved in THF anh. (10 ml THF anh. to 50 mg poly(dichlorophosphazene) and 150 ml THF anh. to remaining poly(dichlorophosphazene)) and $Et_3N$ (0.3 ml to 50 mg, 12 ml to rest) was added prior to the addition of jeffamine-1000 dissolved in 10 ml THF anh. (1 g, 1 mmol to 50 mg) and allylamine (5.4 ml, 71.8 mmol to rest). The jeffamine reaction was stopped after 1 day, and dialysis in water and ethanol was performed to purify the product, yielding 261.6 mg (25 %) of Polymer 1. $^1$H NMR (CDCl$_3$): $\delta$ = 1.11 (s, 9H), 3.37 (s, 3H), 3.63 (s, 85H) ppm. $^{31}$P NMR (CDCl$_3$): $\delta$ = 0.08 ppm. FTIR (solid): $\nu_{max}$ = 2883 (C-H), 1146 (P=N), 1106 (C-O) cm$^{-1}$. The allylamine reaction was stopped after 4 days and the polymer was precipitated from THF in a 50:50 water:ethanol mixture to obtain pure Polymer 2 as a white solid. The reaction mechanisms are shown in Figure 3.4. Yield: 503.9 mg (11 %) $^1$H NMR (CDCl$_3$): $\delta$ = 3.19 (s, 1H), 3.44 (s, 2H), 4.95-5.18 (dd, 2H), 5.81-5.94 (m, 1H) ppm. $^{31}$P NMR (CDCl$_3$): $\delta$ = 2.55 ppm. FTIR (solid): $\nu_{max}$ = 3351 (N-H), 2925 (C-H), 1729 (C=O), 1178 (P=N) cm$^{-1}$.

### 3.3.2 Living polymerization

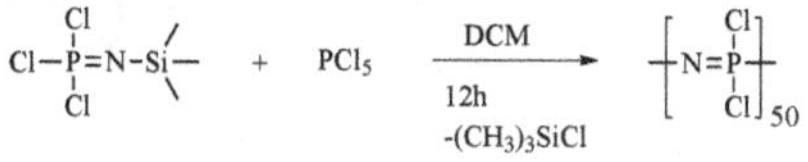

**Figure 3.5** – Living cationic polymerization

According to Allcock et al. [106], the monomer $PCl_3NSi(CH_3)_3$ (2.84 g, 12.7 mmol) was diluted in 10 ml DCM in the glovebox. Then, a solution of $PCl_5$ (0.11 g, 0.5 mmol) in 5 ml DCM was added and the mixture was stirred overnight at room temperature. On the next day, DCM was removed under reduced pressure. The product poly(dichlorophosphazene) was used without any further purification. Yield quantitative. $^{31}$P NMR (CDCl$_3$): $\delta$ = - 18.16 ppm.

### 3.3.3 Macrosubstitution

The poly(dichlorophosphazene) obtained by the living cationic polymerization was further reacted with amines and activated alcohols to get the desired poly(organophosphazene), Polymer 3-5.

**Polymer 3**

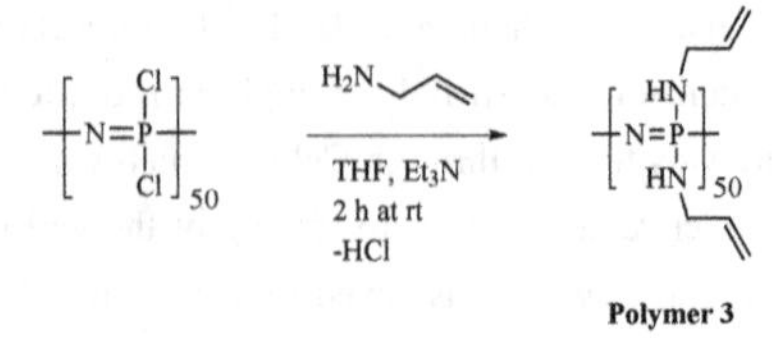

Polymer 3

**Figure 3.6** – Synthesis route for Polymer **3**

The precursor from the living polymerization (1.47 g, 12.65 mmol) was dissolved in a mixture of 10 ml THF anh. and 2 ml Et$_3$N. This precursor solution was slowly added to 2.4 ml allylamine (31.63 mmol) in 40 ml THF anh. and 4.5 ml Et$_3$N. The reaction mixture was stirred for 48 hours at 40 °C under inert conditions. The white precipitate was filtered off and the solvent of the filtrate was removed under reduced pressure. The white precipitate was precipitated in a 1:1 mixture of H$_2$O and ethanol from THF yielding a white precipitate. The reaction mechanism is shown in Figure 3.6. Yield: 1.49 g (74.9 %) $^1$H NMR (CDCl$_3$): $\delta$ = 3.16 (s, 1H), 3.45 (s, 2H), 4.95-5.17 (dd, 2H), 5.81-5.93 (m, 1H) ppm. $^{31}$P NMR (CDCl$_3$): $\delta$ = 2.55 ppm. FTIR (solid): $\nu_{max}$ = 3247 (N-H), 3079-2851 (C-H), 1643 (C=C), 1180 (P=N) cm$^{-1}$.

**Polymer 4**

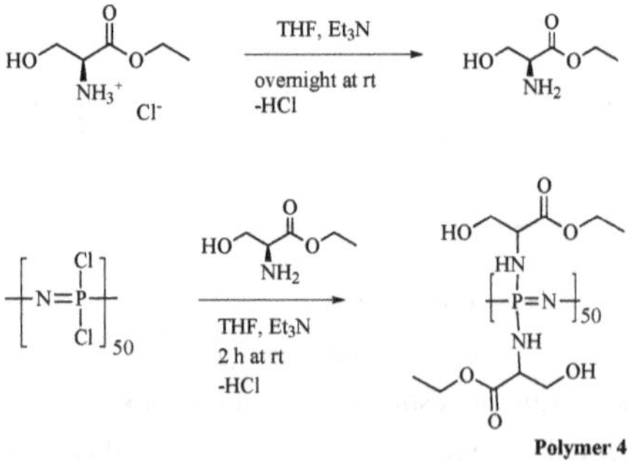

Polymer 4

**Figure 3.7** – Synthesis route for Polymer **4**

According to Morozowich *et al.*, [103] L-Serine ethyl ester (0.57 g, 3.39 mmol) was stirred overnight suspended in 15 ml THF anh. with 1.5 ml Et$_3$N under inert conditions. On the next day, the precursor was dissolved in 10 ml THF anh. with 0.5 ml Et$_3$N and the filtered L-Serine ethyl ester was added. The reaction mixture was stirred for 2 hours at room temperature. The suspension was filtered and solvents were removed under reduced pressure. Then, the polymer was precipitated in petroleum ether, redissolved in THF and, after removing THF under reduced

20

pressure, a yellow oil was obtained. The reaction mechanism is shown in Figure 3.7. Yield: 0.24 g (57.7 %) [1]H NMR (CDCl$_3$): $\delta$ = 1.23-1.35 (m, 3H), 3.88-4.31 (m, 5H) ppm. [31]P NMR (CDCl$_3$): $\delta$ = 1.82 ppm. FTIR (solid): $\nu_{max}$ = 3351 (N-H), 2925 (C-H), 1729 (C=O), 1178 (P=N) cm$^{-1}$.

**Polymer 5**

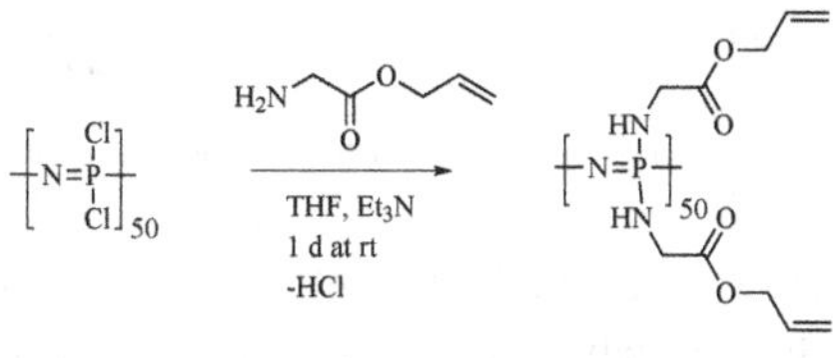

**Figure 3.8** – Synthesis route for Polymer **5**

Poly(dichlorophosphazene) (0.35 g, 2.96 mmol) was dissolved in a mixture of 8 ml THF anh. and 2 ml Et$_3$N and the allyl glycinate (0.82 g, 7.1 mmol) dissolved in 20 ml THF anh. was added. The reaction was stirred overnight in the glovebox. The precipitate Et$_3$N · HCl was removed by filtration, followed by the removal of the solvent under reduced pressure. The polymer was redissolved in ethyl acetate and washed once with water, once with brine and then dried over MgSO$_4$. The pure product was obtained after precipitation in chilled diethyl ether as a highly viscous yellowish oil. The reaction mechanism is shown in Figure 3.8. Yield: 0.65 g (80.5 %) [1]H NMR (CDCl$_3$): $\delta$ = 3.75 (s, 2H), 4.55 (s, 2H), 5.16-5.31 (dd, 2H), 5.82-5.95 (m, 1H) ppm. [31]P NMR (CDCl$_3$): $\delta$ = 1,97 ppm. FTIR (solid): $\nu_{max}$ = 3344 (N-H), 2928 (C-H), 1738 (C=O), 1649 (C=C), 1190 (P=N) cm$^{-1}$.

# 3.4   Thiol-ene photochemistry

## 3.4.1   Functionalization

**Thioglycolic acid**

**Trimer 6**

Trimer **1** (50.3 mg, 0.10 mmol) and 2,2-dimethoxy-2-phenylacetophenone (DMPA) (0.5 mg, 1 wt%) were dissolved in 150 $\mu$l THF. Then thioglycolic acid (14.7 $\mu$l, 0.21 mmol) was mixed into the solution. The reaction mixture was placed for 25 minutes into the UV-chamber and then the solvent was removed yielding Trimer **6** as a highly viscous yellow oil. The reaction mechanism is shown in Figure 3.9. [1]H NMR (CDCl$_3$): $\delta$ = 1.84-1.88 (m, 2H), 2.75-2.76

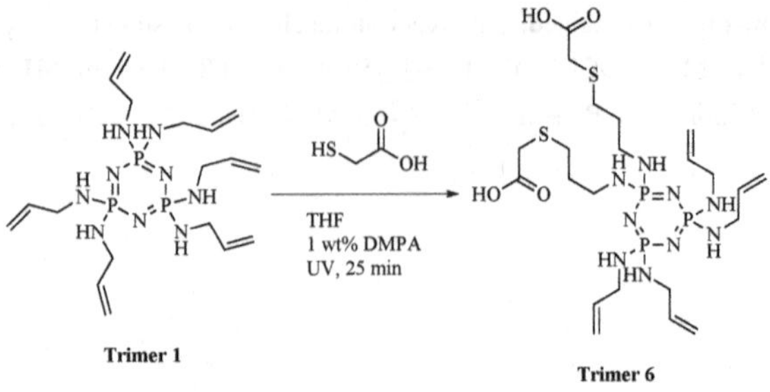

**Figure 3.9** – Synthesis route for Trimer **6** showing one possible isomer.

(m, 2H), 3.07-3.10 (m, 2H), 3.22 (m, 2H), 3.58 (m, 6H), 5.08-5.12 (d, 4H), 5.21-5.29 (m, 4H), 5.83-5.95 (m, 3H) ppm. FTIR (solid): $\nu_{max}$ = 3273 (N-H), 2921-2868 (C-H), 2643 (O-H of COOH) 1704 (C=O), 1645 (C=C), 1239 (P=N) cm$^{-1}$.

**Trimer 7**

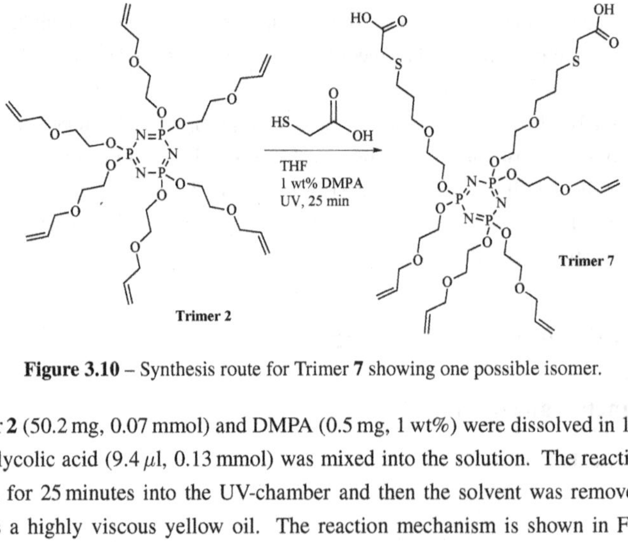

**Figure 3.10** – Synthesis route for Trimer **7** showing one possible isomer.

Trimer **2** (50.2 mg, 0.07 mmol) and DMPA (0.5 mg, 1 wt%) were dissolved in 150 μl THF. Then thioglycolic acid (9.4 μl, 0.13 mmol) was mixed into the solution. The reaction mixture was placed for 25 minutes into the UV-chamber and then the solvent was removed yielding Trimer **7** as a highly viscous yellow oil. The reaction mechanism is shown in Figure 3.10. $^{1}$H NMR (CDCl$_3$): $\delta$ = 1.85-1.90 (t, 2H), 2.77-2.81 (t, 2H), 3.26 (s, 2H), 3.58-3.63 (m, 2H), 3.66-3.69 (m, 4H), 4.04-4.11 (m, 6H), 5.18-5.33 (m, 2H), 5.84-5.97 (m, 1H) ppm. FTIR (solid): $\nu_{max}$ = 2923 (C-H), 2644 (O-H of COOH), 1729 (C=O), 1646 (C=C), 1221 (P=N) cm$^{-1}$.

## Ethyl-3-mercaptopropionate

## Trimer 8

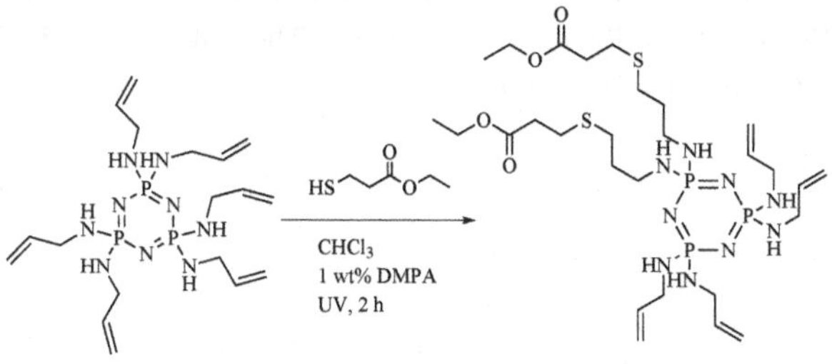

**Figure 3.11** – Synthesis route for Trimer **8** showing one possible isomer.

Trimer **1** (50.8 mg, 0.11 mmol) and DMPA (0.5 mg, 1 wt%) were dissolved in 150 $\mu$l CHCl$_3$. Then ethyl-3-mercaptopropionate (26.9 $\mu$l, 0.21 mmol) was mixed into the solution. The reaction mixture was placed for 30 minutes into the UV-chamber and the solvent was removed yielding Trimer **8** as a highly viscous yellow oil. The reaction mechanism is shown in Figure 3.11. $^1$H NMR (CDCl$_3$): $\delta$ = 1.24-1.29 (t, 6H), 1.73-1.82 (m, 4H), 2.56-2.62 (m, 8H), 2.73-2.79 (m, 4H), 3.01 (s, 4H), 3.56 (s, 8H), 4.12-4.21 (m, 4H), 5.04-5.08 (td, 4H), 5.19-5.26 (qd, 4H), 5.85-5.98 (m, 4H) ppm. FTIR (solid): $\nu_{max}$ = 3230 (N-H), 2981-2919 (C-H), 1731 (C=O), 1643 (C=C), 1180 (P=N) cm$^{-1}$.

## Glutathione

## Trimer 9

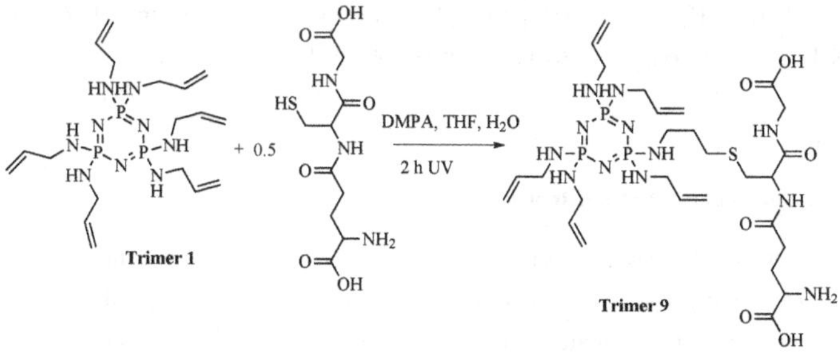

**Figure 3.12** – Synthesis route for Trimer **9**

Glutathione (48.9 mg, 0.16 mmol) dissolved in 800 $\mu$l H$_2$O and Trimer **1** (50.0 mg, 0.32 mmol) with DMPA (0.5 mg, 1 wt%) dissolved in 800 $\mu$l THF were mixed. The mixture

23

was placed for 2 h into the UV-chamber yielding a turbid solution. The solvents were removed under reduced pressure to obtain a white precipitate. The reaction mechanism is shown in Figure 3.12. $^1$H NMR (D$_2$O): $\delta$ = 1.71-1.76 (t, 1H), 2.07-2.14 (q, 1H), 2.46-2.60 (td, 2H), 2.77-3.04 (m, 2H), 3.49 (s, 2H), 3.70-3.75 (t, 1H), 3.83 (s, 1H), 4.50-4.54 (t, 0.5H), 5.07-5.24 (dd, 2H), 5.81-5.93 (m, 1H) ppm. $^{31}$P NMR (D$_2$O): $\delta$ = 13.6 ppm. FTIR (solid): $\nu_{max}$ = 3273 (N-H), 2921 (C-H), 1704 (C=O), 1645 (C=C), 1239 (P=N) cm$^{-1}$.

**Polymer 6**

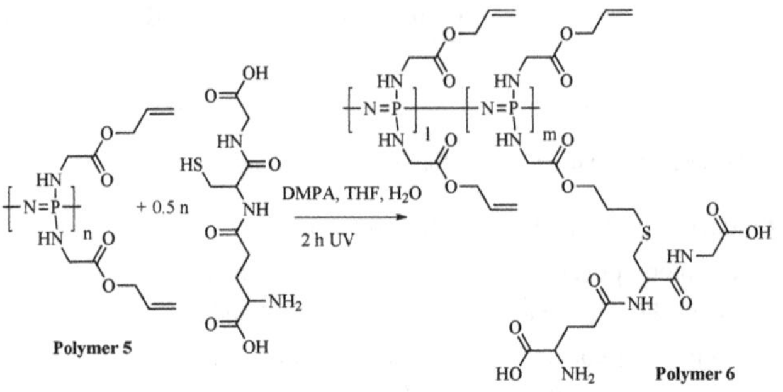

**Figure 3.13** – Synthesis route for Polymer **6**

Glutathione (45.0 mg, 0.15 mmol) dissolved in 800 $\mu$l H$_2$O and Polymer **5** (80.0 mg, 0.29 mmol) with DMPA (0.8 mg, 1 wt%) dissolved in 800 $\mu$l THF were mixed. The mixture was placed for 2 h into the UV-chamber yielding a turbid solution. The solvents were removed under reduced pressure to obtain a white precipitate. The reaction mechanism is shown in Figure 3.13. $^1$H NMR (D$_2$O acidified with TFA): $\delta$ = 0.68 (s, 0.5H), 1.06 (s, 0.7H), 1.43 (s, 1.3H), 1.64-1.72 (m, 0.6H), 1.91 (s, 0.1H), 2.44 (s, 0.2H), 2.69-3.07 (m, 4.4H), 3.37 (s, 1.7H), 3.44 (m, 0.4H), 4.01 (m, 2H), 4.62 (s, 1H) ppm. $^{31}$P NMR (D$_2$O): $\delta$ = 4.11 ppm. FTIR (solid): $\nu_{max}$ = 3256 (N-H), 2923 (C-H), 1732 (C=O), 1195 (P=N) cm$^{-1}$.

**Fluorescein**

**Fluorescein isothiocyanate modification**

Using a method adapted from Kaufmann et al., [107] (tritylsulfanyl)-ethylamine (164.1 mg, 0.51 mmol) was dissolved in 8 ml DCM under water free conditions and cooled to 0 °C. Fluorescein isothiocyanate isomer I (200 mg, 0.51 mmol) was dissolved in 16 ml of a 1:1 mixture of DCM:DMF and carefully added to the stirred (tritylsulfanyl)ethylamine solution. The mixture was allowed to stir overnight at room temperature. On the next day, the solvent was removed under reduced pressure yielding an orange precipitate.

24

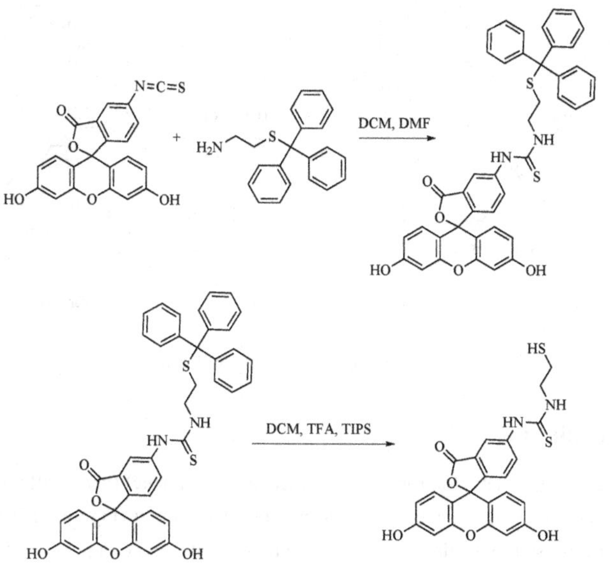

**Figure 3.14** – Fluorescein isothiocyanate modification

The deprotection was performed using a mixture of DCM:TFA:TIPS (45:50:5, 9:10:1 ml). The orange powder was dissolved in DCM and then TFA and TIPS were added. The reaction mixture was stirred for 1 hour at room temperature. Then, the reaction was stopped by removing the solvents under reduced pressure. The orange precipitate was recrystallized in diethylether in the freezer overnight yielding an orange precipitate, which was used without further purification. The reaction mechanism is shown in Figure 3.14. Yield: 220.07 mg (91.8 %)

**Polymer 7**

Polymer 5 (150 mg, 0.55 mmol), modified fluorescein (5.1 mg, 0.01 mmol, 2 mol%) and DMPA (1.5 mg, 1 wt%) were dissolved in 3 ml $CHCl_3$. The mixture was shaken thoroughly and placed in the UV-chamber for 15 minutes. The product was used immediately for crosslinking without further purification. The reaction mechanism is shown in Figure 3.15.

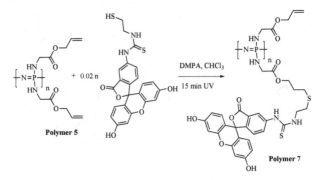

Figure 3.15 – Synthesis route for Polymer **7**

## 3.4.2 Crosslinking

The reaction of trimethylolpropane tris(3-mercaptopropionate) (herein after referred to as trithiol) with the following organic molecules (Figure 3.16), trimers (Figure 3.17) and polymers (Figure 3.18) was used to obtain a 3-dimensional network in form of a pellet with and without pores.

**Organic molecules used for crosslinking**

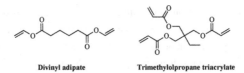

Divinyl adipate        Trimethylolpropane triacrylate

Figure 3.16 – Organic molecules used for crosslinking

**Trimers used for crosslinking**

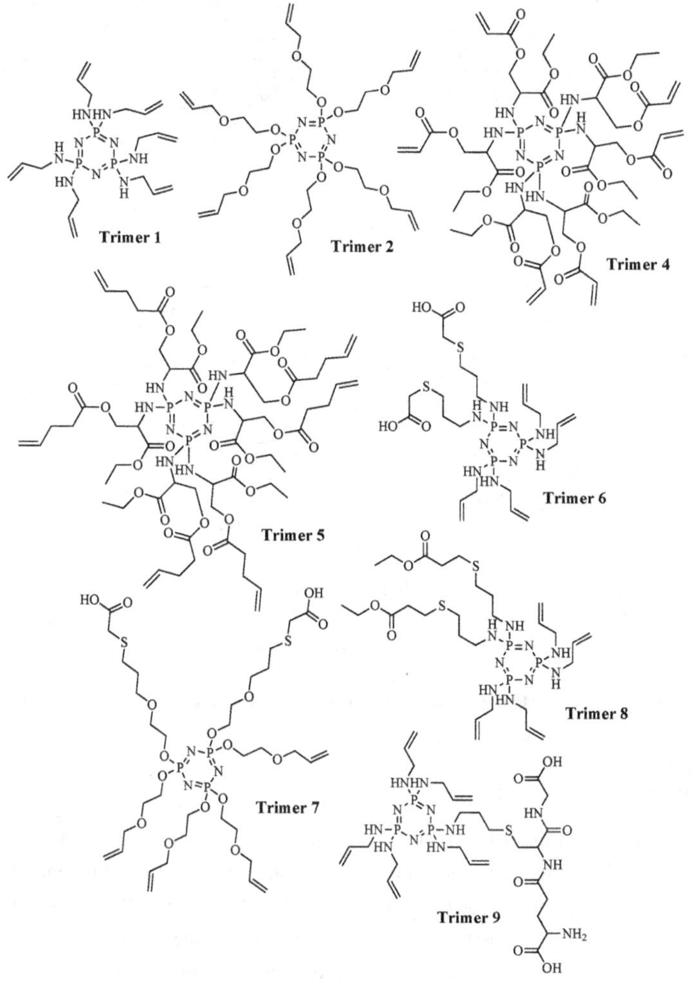

**Figure 3.17** – Trimers used for crosslinking

**Polymers used for crosslinking**

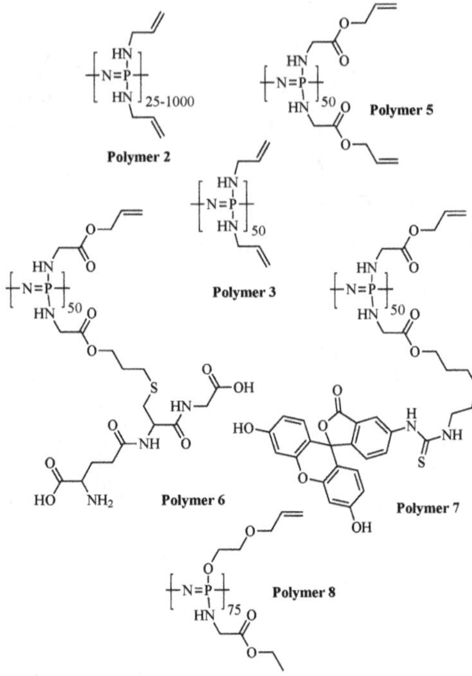

**Figure 3.18** – Polymers used for crosslinking

**Procedure**

The conditions for the thiol-ene crosslinking reactions were all similar to the reaction described below using different molar ratios of thiol to alkene groups and different amounts of solvent. The exact quantities used are given in Table 3.1, in which "Name" stands for the trimer/polymer used for the crosslinking reaction, "Amount" for the amount of this trimer/polymer, "Solvent" for the solvent and the amount of it used, "Trithiol" for the percentage of crosslinking and the amount used, "Triacrylate" for trimethylolpropane triacrylate, for the percentage of double bonds added and the amount used, "Divinyl adipate" for the percentage of double bonds added and the amount used and "Number" for the new number of the trimer/polymer. This "Number" consists of number-c-number, whereby the first number stands for the trimer/polymer used, the "c" stands for "crosslinked" and the second number for counting e.g. Trimer **1c3** is the third crosslinked network made from Trimer **1**. As an example the procedure for Trimer **1c1** is given:

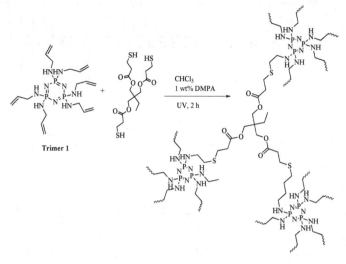

**Figure 3.19** – Crosslinking Trimer **1c1**

Trimer **1** (50.2 mg, 0.10 mmol) and DMPA (0.5 mg, 1 wt%) were dissolved in 150 $\mu$l CHCl$_3$. Trithiol (70 $\mu$l, 0.21 mmol) was added and the mixture was shaken thoroughly, before it was placed for 1 hour in the UV-chamber. The resulting material was a brittle yellow clear pellet. FTIR (solid): $\nu_{max}$ = 3357 (N-H), 2929 (C-H), 1728 (C=O), 1181 (P=N) cm$^{-1}$.

**Table 3.1** – Conditions for crosslinking

| Name | Amount [mg, mmol polymer] | Solvent | Solvent [µl] | Trithiol [%, µl] | Triacrylate [% db, µl]ᵃ | Divinyl adipate [% db, µl] | Number |
|---|---|---|---|---|---|---|---|
| Trimer 1 | 50, 0.11 | CHCl$_3$ | 150 | 100, 69.9 | 0, 0 | 0, 0 | 1c1 |
| Trimer 1 | 50, 0.11 | CHCl$_3$ | 150 | 75, 52.4 | 0, 0 | 0, 0 | 1c2 |
| Trimer 1 | 50, 0.11 | CHCl$_3$ | 150 | 50, 34.9 | 0, 0 | 0, 0 | 1c3 |
| Trimer 1 | 50, 0.11 | CHCl$_3$ | 150 | 25, 17.5 | 0, 0 | 0, 0 | 1c4 |
| Trimer 2 | 30, 0.04 | CHCl$_3$, EtOH | 135, 405 | 100, 26.6 | 0, 0 | 0, 0 | 2c1 |
| Trimer 2 | 30, 0.04 | CHCl$_3$, EtOH | 135, 405 | 75, 20.0 | 0, 0 | 0, 0 | 2c2 |
| Trimer 2 | 30, 0.04 | CHCl$_3$, EtOH | 135, 405 | 50, 13.3 | 0, 0 | 0, 0 | 2c3 |
| Trimer 4 | 33, 0.03 | CHCl$_3$ | 150 | 100, 17.6 | 0, 0 | 0, 0 | 4c1 |
| Trimer 5 | 30, 0.02 | CHCl$_3$ | 500 | 100, 13.9 | 0, 0 | 0, 0 | 5c1 |
| Trimer 5 | 30, 0.02 | CHCl$_3$, MeOH | 25, 600 | 100, 13.9 | 0, 0 | 0, 0 | 5c2 |
| Trimer 5 | 30, 0.02 | DMSO | 500 | 100, 13.9 | 0, 0 | 0, 0 | 5c3 |
| Trimer 5 | 30, 0.02 | AcN | 500 | 100, 13.9 | 0, 0 | 0, 0 | 5c4 |
| Trimer 6 | 70, 0.11 | THF | 150 | 100, 46.6 | 0, 0 | 0, 0 | 6c1 |
| Trimer 6 | 70, 0.11 | THF | 150 | 75, 35.0 | 0, 0 | 0, 0 | 6c2 |
| Trimer 7 | 59, 0.07 | THF | 150 | 75, 22.2 | 0, 0 | 0, 0 | 7c1 |
| Trimer 8 | 79, 0.11 | CHCl$_3$ | 150 | 100, 46.6 | 0, 0 | 0, 0 | 8c1 |
| Trimer 8 | 79, 0.11 | CHCl$_3$ | 150 | 75, 35 | 0, 0 | 0, 0 | 8c2 |
| Trimer 9 | 47, 0.05 | THF, H$_2$O | 500, 500 | 100, 24.6 | 0, 0 | 0, 0 | 9c1 |
| Polymer 2 | 30, 0.19 | THF | 150 | 100, 41.9 | 0, 0 | 0, 0 | 2c1 |
| Polymer 3 | 400, 02.55 | CHCl$_3$ | 1000 | 100, 559.2 | 0, 0 | 0, 0 | 3c1 |
| Polymer 5 | 30, 0.11 | CHCl$_3$ | 150 | 100, 24.1 | 0, 0 | 0, 0 | 5c1 |
| Polymer 5 | 10, 0.04 | CHCl$_3$ | 150 | 100, 40.3 | 0, 0 | 80, 27.7 | 5c2 |
| Polymer 6 | 78, 0.18 | THF, H$_2$O | 500, 500 | 100, 30.1 | 0, 0 | 0, 0 | 6c1 |
| Polymer 8 | 30, 0.12 | CHCl$_3$ | 500 | 100, 13.2 | 0, 0 | 0, 0 | 8c1 |
| Polymer 8 | 30, 0.12 | CHCl$_3$ | 500 | 75, 9.9 | 0, 0 | 0, 0 | 8c2 |
| Polymer 8 | 10, 0.04 | CHCl$_3$ | 500 | 100, 39.6 | 0, 0 | 89, 30.3 | 8c3 |
| Polymer 8 | 10, 0.04 | CHCl$_3$ | 500 | 100, 39.6 | 89, 28.7 | 0, 0 | 8c4 |
| Test substance | - | CHCl$_3$, EtOH | 135, 405 | 100, 33.4 | 100, 27.3 | 0, 0 | 1 |
| Test substance | - | CHCl$_3$ | 150 | 100, 55.4 | 0, 0 | 100, 47.6 | 2 |

ᵃ% db = percent of double bonds reacted with trithiol

## 3.5 Pore generation - matrix formation

The composition of the reaction mixtures was similar to those used for crosslinking. NaCl was added as a porogen and PEG-200 as lubricant to enable the removal of the pellets from the glass vials. The conditions especially in terms of amount of solvent and lubricant for the pore formation vary slightly between the different trimers and polymers. Table 3.2 gives the amounts determined to lead to the best results. The "Number" here consists of number-p-number, whereby the first number stands for the trimer/polymer used to produce the matrix, the "p" stands for "porous" and the second number for counting. For instance, Polymer **5p2** is the second porous network made from Polymer **5**. The procedure is similar in all cases and works as described below for Polymer **5c4**:

Polymer **5** (60 mg, 0.22 mmol) and DMPA (0.6 mg, 1 wt%) were dissolved in 0.5 ml CHCl$_3$ and 0.5 ml PEG-200 were added as lubricant. Trithiol (48.3 $\mu$l, 0.15 mmol) was added and the mixture was shaken thoroughly. Then about 4 g of NaCl were immersed in the crosslinking solution. The sample was placed for 2 hours in the UV-chamber. The white-yellowish pellet was kept in a water bath for 1 day to wash out the salt and PEG-200. Then it was cleaned with a Soxhlet extraction with ethanol for 2 days and afterwards dried in a vacuum drying oven at 40 °C leading to a porous pellet. FTIR (solid): $\nu_{max}$ = 3351 (N-H), 2925 (C-H), 1729 (C=O), 1178 (P=N) cm$^{-1}$.

For the pellets with Trimer **1** also some trials with PEG-1000 and PEG-2000 dissolved in THF instead of PEG-200 were performed, but no differnce to PEG-200 was observed. Therefore only PEG-200 was used for the further experiments.

**Table 3.2** – Conditions for pore generation

| Name | Amount [mg, mmol polymer] | Solvent | Solvent [ml] | PEG-200 [ml] | Trithiol [%, μl] | Triacrylate [% db, μl]$^a$ | Divinyl adipate [% db, μl] | Number |
|---|---|---|---|---|---|---|---|---|
| Trimer 1 | 100, 0.21 | THF | 1 | 1.5 | 100, 140.0 | 0, 0 | 0, 0 | 1p1 |
| Trimer 1 | 50, 0.11 | THF | 1 | 1.5 | 100, 140.0 | 50, 57.1 | 0, 0 | 1p2 |
| Trimer 2 | 100, 0.14 | THF | 1 | 0.3 | 75, 66.6 | 0, 0 | 0, 0 | 2p1 |
| Trimer 2 | 50, 0.07 | THF | 1 | 0.5 | 100, 88.8 | 50, 36.3 | 0, 0 | 2p2 |
| Polymer 3 | 100, 0.64 | THF | 1 | 1 | 100, 140.0 | 0, 0 | 0, 0 | 3p1 |
| Polymer 3 | 50, 0.32 | THF | 1 | 1 | 100, 140.0 | 50, 57.11 | 0, 0 | 3p2 |
| Polymer 3 | 50, 0.32 | THF | 1 | 1 | 100, 140.0 | 0, 0 | 50, 60.0 | 3p3 |
| Polymer 5 | 60, 0.22 | CHCl$_3$ | 0.5 | 0.5 | 100, 48.3 | 0, 0 | 0, 0 | 5p1 |
| Polymer 5 | 60, 0.22 | CHCl$_3$ | 0.5 | 0.5 | 100, 60.3 | 0, 0 | 20, 10.4 | 5p2 |
| Polymer 5 | 50, 0.18 | CHCl$_3$ | 0.5 | 0.5 | 100, 80.5 | 0, 0 | 50, 34.6 | 5p3 |
| Polymer 5 | 20, 0.07 | CHCl$_3$ | 0.5 | 0.5 | 100, 80.5 | 0, 0 | 80, 55.4 | 5p4 |
| Polymer 7 | 62, 0.22 | CHCl$_3$ | 0.5 | 0.5 | 100, 48.3 | 0, 0 | 0, 0 | 7p1 |
| Polymer 7 | 52, 0.18 | CHCl$_3$ | 1 | 0.5 | 100, 80.4 | 0, 0 | 50, 34.6 | 7p2 |
| Polymer 8 | 20, 0.08 | CHCl$_3$ | 0.6 | 0.4 | 100, 79.3 | 0, 0 | 89, 60.6 | 8p1 |
| Test substance | - | CHCl$_3$ | 0.5 | 0.5 | 100, 110.8 | 0, 0 | 100, 95.2 | 3 |

$^a$% db = percent of double bonds reacted with trithiol

# 3.6 Degradation studies

### 3.6.1 Trimer 1c1 and Polymer 3c1

To determine the gravimetric mass loss, 50 mg of Trimer **1c1**/Polymer **3c1** were placed in sealed vials and incubated at 37 °C in either 2 ml Tris-buffer (pH 7.4) or 2 ml deionized $H_2O$ + HCl (pH 2). Always, one sample per pH was taken out at each time point. The degradation medium was removed. The samples were washed and dried in a vaccuum drying oven at 40 °C until the weight was constant.

### 3.6.2 Polymer 3p1, 3p2 and 3p3

The gravimetric mass loss was determined by placing 30 mg of Polymer **3p1**, **3p2** and **3p3** in sealed vials and incubating them at 37 °C in either 2 ml Tris-buffer (pH 7.4) or 2 ml deionized $H_2O$ + HCl (pH 2). Always, one sample per pH was taken out at each time point. The degradation medium was removed. The samples were washed and dried in a vaccuum drying oven at 40 °C until the weight was constant.

### 3.6.3 Polymer 5p1, 5p3 and Test substance 3

The mass loss was determined gravimetrically by placing 30 mg of Polymer **5p1**, **5p3** or Test substance 3 in sealed vials and incubating them at 37 °C in 2 ml deionized $H_2O$ (pH 6.2). At every time point, one sample was measured. The degradation medium was removed. The samples were washed and dried in a vaccuum drying oven at 40 °C until the weight was constant.

# 4. Results and discussion

## 4.1 Trimers

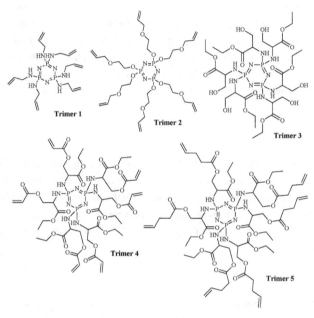

**Figure 4.1** – Trimers, which were investigated in this thesis.

**Trimer 1**

Trimer **1** was used as a model compound. Most experiments were first performed using this trimer as it was easy in handling and cheap to produce. Besides, it is very reactive, thus performing reactions with Trimer **1** gave an idea about the potential of the thiol-ene chemistry using poly(ogranophosphazenes). Unfortunately, this trimer is toxic, not degradable and hydrophobic and hence not applicable for cell seeding.

**Trimer 2**

Ethylene glycol allyl ether was used as a side chain for Trimer 2 as it is more hydrophilic and less toxic than allylamine. Trimer 2 was synthesized in a very good yield and quite pure. Although the trimer was not completely soluble in the used solvents, all reactions could be performed.

**Trimer 3**

Serine as a side chain does not just give a very hydrophilic trimer, but also allows modifications on the free OH-group. Thus, for instance double bonds via acrylates (Trimer 4) or anhydrides (Trimer 5) may be added to create a versatile system. Moreover, the synthesis of Trimer 3 led to several side-products and impurities. Therefore, the raw product was purified by dry column vacuum chromatography giving the product, but in a very low yield and not perfectly pure. Hence, the production of this trimer was not feasible and was not carried out further.

**Trimer 4**

Trimer 4 with an acrylate modified L-Serine should be more hydrophilic than Trimer 1 and 2. Unfortunately, the product showed various side products and impurities. Purification with dry column vacuum chromatography gave a small amount of the pure product as an orange oil. Thus the concept works in principle, however many acrylates are harmful for living organisms [108], therefore no further investigations in the production of the L-Serine trimer with the acrylate modification were carried out.

**Trimer 5**

The side chain of Trimer 5 was made of a L-Serine modified with pentenoic anhydride at the OH-group. This should give a hydrophilic and biocompatible trimer. The production of this trimer was very laborious and as the follow up experiments did not give the desired results, also the production of this trimer was discontinued.

## 4.1.1 Crosslinking

A radical reaction between double bonds of the trimers/polymers and thiol groups of trithiol was used to obtain a crosslinked network with infinite molecular weight. A schematic drawing of the crosslinked network produced is shown in Figure 4.2. Figure 4.3 shows the FTIR measurements of Trimer 1 and the crosslinked Trimer 1c1. The dissappearance of the double bond band and the appearance of the ester band in Trimer 1c1 can clearly be seen. Generally, crosslinking gives the polymers stability and form to allow the material in the later course to

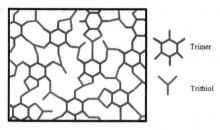

Trimer

Trithiol

**Figure 4.2** – Simplified scheme of the network formed by crosslinking a trimer with trithiol

carry cells without collapsing. The different polymers and trimers possess different properties and also vary tremendously in the final crosslinked stage, from very brittle to rather gel-like structures. Also, the ratio of thiol-groups to double bonds plays an important role in the final appearance. Generally fully crosslinked materials, so a ratio of 1:1 thiol:double bond, were favored by the cells, as shown in Chapter 4.7.

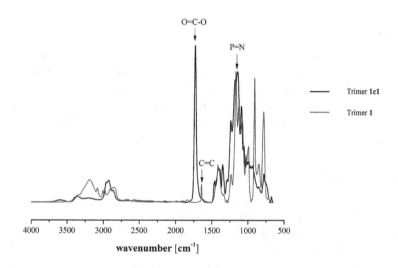

**Figure 4.3** – FTIR comparison of Trimer **1c1** and Trimer **1**. The typical band for carbon double bonds at $1644\,\text{cm}^{-1}$ disappears and the band for carbon oxygen bonds of esters around $1730\,\text{cm}^{-1}$ appears in the crosslinked Trimer **1c1**.

**Trimer 1**

Trimer **1** was crosslinked with 100, 75, 50 and 25 % of the double bonds used by trithiol giving Trimer **1c1**, **1c2**, **1c3** and **1c4** as clear yellowish pellets. The first three were quite brittle, whereas the last one was rather gel-like. FTIR measurements were performed to observe the change after crosslinking as shown in Figure 4.3. Trimer **1c1** was prepared for degradation studies. Trimer **1c1**, **1c2** and **1c3** were also prepared as non-transparent thin films in 48-well plates and as powdered materials for biomedical testing.

**Trimer 2**

This trimer was crosslinked with 100, 75 and 50 % of the double bonds used by trithiol in 48-well plates to obtain thin films for biomedical testing. Trimer **2c1**, **2c2** and **2c3** formed non-transparent films like Trimer **1c1**, **1c2** and **1c3**, therefore making several common biological tests impossible.

**Trimer 4**

The small amount of Trimer **4** obtained was enough for one crosslinking experiment only giving Trimer **4c1**. Thus, it is evident, that a modification and later crosslinking is possible, but not practicable.

**Trimer 5**

Trimer **5** was used for crosslinking experiments. The main challenge was to dissolve the trimer. Methanol was the ideal solvent, but disturbed UV-crosslinking. The other solvents could not dissolve the trimer enough to enable a full reaction with trithiol. A reduction of double bonds was observed in the FTIR spectra, but the reaction was not complete, so no pellets of Trimer **5c1**, **5c2**, **5c3** and **5c4** were produced. Thus, further investigations in Trimer **5** were not performed.

### 4.1.2   Pore generation

The porous structures were obtained using sodium chloride as a porogen. This salt is perfectly suited for UV-photocrosslinkage as it can be penetrated by UV-light allowing the photoreactions to take place. It is easily washed out, besides sodium chloride is not toxic, so the cells would not be harmed by left traces. The porous materials obtained also varied in terms of appearance and mechanical properties. While the pure trimer/polymer pellets were quite hard and stiff, blending with divinyl adipate made the pellets softer, more flexible, but less elastic. All pellets were not brittle at all and flexible enough to cut them into shape - a beneficial property for tissue engineering applications. The pore size distribution showed that most pores have a diameter above 70 $\mu$m and lie between 70 and 250 $\mu$m. Table 4.1 shows the mean average pore volume and diameter and the number of pores of Trimer **1p1** and **2p1** and Polymer **3p1**, **5p1** and **5p4** determined by CT on a block with the dimensions 2490 x 1268 x 702 $\mu$m. Additionally, highly interconnected networks were produced. Thus, the infiltration of the material with various cell types and the communication between embedded cells is possible. In other words the shape of the matrices is suitable for tissue engineering.

**Table 4.1** – CT results of Trimer **1p1** and **2p1** and Polymer **3p1**, **5p1** and **5p4**

| Name | Average pore volume [$\mu m^3$] | Average mean pore diameter [$\mu m$] | Number of pores |
|---|---|---|---|
| Trimer **1p1** | $2.23 \cdot 10^6$ | 171.2 | 304 |
| Trimer **2p1** | $1.11 \cdot 10^6$ | 129.1 | 577 |
| Polymer **3p1** | $3.22 \cdot 10^6$ | 202.3 | 213 |
| Polymer **5p1** | $2.07 \cdot 10^6$ | 167.8 | 378 |
| Polymer **5p4** | $2.70 \cdot 10^6$ | 144.0 | 481 |

**Trimer 1**

Porous pellets were fabricated using Trimer **1** giving Trimer **1p1** as a robust, but not brittle pellet. In Figure 4.4 two SEM pictures of different positions on a pellet of Trimer **1p1** are shown. Figure 4.5 and Table 4.1 show the results of the CT measurement of Trimer **1p1**. These SEM and CT results confirm a highly interconnected porous structure of the pellets. The average pore volume exceeds typical eukaryotic cell volume, hence the matrix shape is ideally suitable for tissue engineering. Unfortunately, cell seeding tests could not be performed with Trimer **1p1** as it was too hydrophobic. For this reason more hydrophilic side chains were used in the later course.

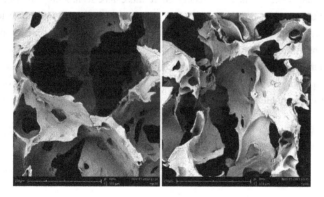

**Figure 4.4** – SEM pictures of Trimer **1p1**. Scale bars equal 250 $\mu m$

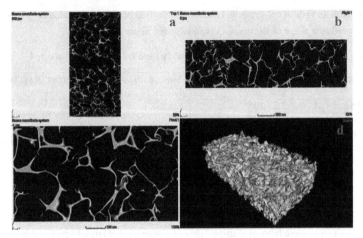

**Figure 4.5** – CT pictures of Trimer **1p1**. 2D View from top, scale bar equals 400 $\mu$m (a), from right side, scale bar equals 250 $\mu$m (b) and from front side, scale bar equals 150 $\mu$m (c). 3D View of the block (d)

**Trimer 2**

Porous material was produced from Trimer **2** namely Trimer **2p1**, which appeared less stiff and more flexible than Trimer **1p1**. Figure 4.6 shows the SEM images of Trimer **2p1**. In Figure 4.7 and Table 4.1 the CT measurement results can be seen. Although Trimer **2p1** had the smallest pores among the tested samples, it still possessed a highly interconnected porous matrix, showing suitable dimensions for cell seeding. Trimer **2p1** was more hydrophilic than Trimer **1p1**, but it was still too hydrophobic for the Red Cross team to perform cell seeding tests. BioMed-zet managed to conduct a short cell seeding cytotoxicity test. Therefore, the focus was directed on poly(organophosphazenes) with amino acid containing side chains to obtain more hydrophilic and more biocompatible matrices.

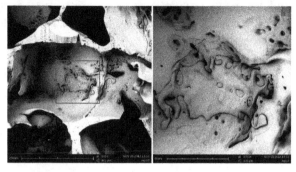

**Figure 4.6** – SEM picture of Trimer **2p1**. Scale bar equals 220 $\mu$m (a), rectangle in (a) further magnified so that scale bar equals 60 $\mu$m (b)

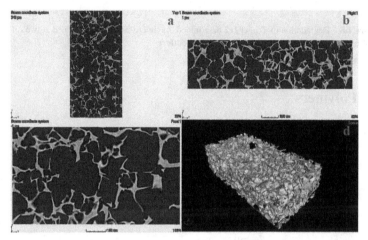

**Figure 4.7** – CT picture of Trimer **2p1**. 2D View from top, scale bar equals 400 $\mu$m (a), from right side, scale bar equals 250 $\mu$m (b) and from front side, scale bar equals 150 $\mu$m (c). 3D View of the block (d)

### 4.1.3   Degradation studies

**Trimer 1c1**

Mass loss experiments were performed with Trimer **1c1** over a time of 21 weeks at two pH values namely pH 2 for enhanced conditions and pH 7.4 for physiological conditions. In general, the degradation of the network was very slow, as expected for trimer based materials and a significant difference between the two pH values could not be observed as shown in Figure 4.8. Both test series of Trimer **1c1** showed a mass loss of approximately 10 % after 21 weeks. ASC are able to synthesize cartilage matrix within 4-5 weeks [109], therefore the observed degradation rate of the matrix is far too low for tissue engineering applications. Additionally, allylamine is toxic, so probably toxic degradation products would be released, giving another reason for not continuing with this type of network.

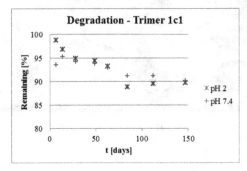

**Figure 4.8** – Degradation of Trimer **1c1** determined gravimetrically and performed at two different pH values, namely, 2 (HCl in $H_2O$) and 7.4 (Tris-buffer)

## 4.2 Polymers

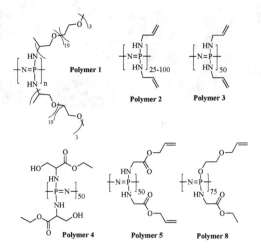

**Figure 4.9** – Polymers, which were investigated in this thesis.

**One-pot polymerization**

The disadvantage of the living polymerization is the demanding monomer production including distillation steps leading to high losses. Furthermore, high molecular weight polymers are not yet feasable with this method. For this reason, a one-pot synthesis was performed with the objective to produce a high molecular weight polymer in a single reaction row without any purification step during the delicate poly(dichlorophosphazene) production.

**Polymer 1**

The polymer produced by this method was substituted with jeffamine and obtained as a pure wax after dialysis. The yield can certainly be improved by optimizing the technique. Jeffamine was chosen as a side chain to be able to determine the number of repetition units and hence the size of the polymer using GPC. The GPC-measurement performed confirmed that a high molecular weight polymer was produced as shown in Figure 4.10. Interestingly, two distinct peaks were observed in the GPC-measurement, one at 9.31 suggesting about 25 repetition units and another one at 6.56 suggesting about 100 repetition units [46]. It is unclear, if the smaller polymer is a kind of side-product, a degradation product or something completely different. The $^{31}$P NMR showed just one single broad peak, proving that the two kind of chains are identical in terms of chemical composition.

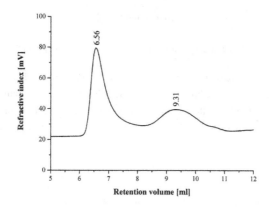

**Figure 4.10** – GPC measurement of Polymer 1

**Polymer 2**

An allylamine polymer was also produced by this technique, since allylamine is quite easy in handling, it is a small side chain and has known properties. The pure product was obtained

43

in a low yield due to some challenges in the purification steps. It was not possible to perform a GPC measurement with this polymer, but when it was dissolved in THF, the solution was quite viscous, also giving a clue, that it has a very high molecular weight.

**Living polymerization**

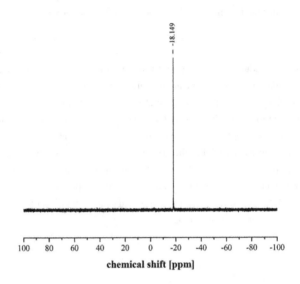

chemical shift [ppm]

**Figure 4.11** – [31]P NMR of the precursor poly(dichlorophosphazene) from Polymer 5.

The significant advantages of living polymerization are the narrow polydispersity and the control over chain length. The method for this cationic polymerization was optimized by laboratory members [110]. Therefore, the poly(dichlorophosphazene) production with 50 repetition units always lead to a quantitative yield and a single species in [31]P NMR as shown in Figure 4.11.

**Polymer 3**

The simple Polymer 3 with allylamine side chains was prepared as a model polymer and several tests were first performed using this polymer. Hence, it was used for several different crosslinking experiments and as a basis for degradations studies.

**Polymer 4**

The amino acid L-serine as a side chain makes a polymer more hydrophilic than for instance allylamine. Additionally, serine can be modified via the OH-group. Unfortunately, the

44

NMR of Polymer 4 showed some side-products and impurities. It was impossible to free the polymer from all impurities like BHT stemming from THF, although several purification steps were performed. Later trials by laboratory colleagues using dialysis failed as well. Therefore, also the production of the serine polymer was not feasible. For this reason, it was decided to combine the amino acid and the double bond donor before the addition to the polymer.

**Polymer 5**

Polymer 5 with allyl glycinate as a side chain was obtained as a pure product in a very good yield. The considerable improvement achieved by this polymer is, that it contains a combination of an amino acid and a double bond donor. Therefore, Polymer 5 is more hydrophilic and should degrade faster than for instance Polymer 3 [111]. The amino acid additionally guarantees the biocompatibility and hence, the suitability of Polymer 5 for the tissue engineering project. Therefore, Polymer 5 was chosen as the ideal polymer for future work.

**Polymer 8**

The mixed Polymer 8 contained a glycine ester and an ethylene glycol allyl ether to fulfill the requirement of a hydrophilic and fast degrading polymer via the glycine ester [111] and a non-toxic double bond donor via the ethylene glycol allyl ether. The significant disadvantage of mixed polymers is that it is impossible to repeat the exact ratio of the two side chains. Thus, this kind of polymers are interesting for chemical, mechanical and biological tests, but in the end inapplicable for this project.

## 4.2.1 Crosslinking

The principle behind crosslinking polymers is the same as with trimers. Figure 4.12 shows a schematic drawing of a crosslinked polymer.

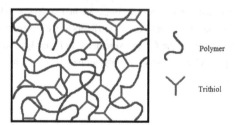

**Figure 4.12** – Simplified scheme of the network formed by crosslinking a polymer with trithiol

**Polymer 2**

Polymer 2 was crosslinked 100 % to be able to check, whether crosslinking is possible and to have a comparison to Polymer 3c1. The resulting Polymer 2c1 was softer and less brittle than Polymer 3c1, so obviously, increasing the molecular weight from approximately 13 660 to 27 321 g/mol (assuming 50 and 100 repetition units) has a tremendous impact on the final polymer properties.

**Polymer 3**

Polymer 3 was 100 % crosslinked giving Polymer 3c1. This very brittle material was further used for degradation studies to compare the degradation of the allylamine trimer with the polymer.

**Polymer 5**

A crosslinking experiment of 100 % pure Polymer 5 was performed. The reaction lead to a nicely shaped pellet named Polymer 5c1.

**Polymer 8**

Several crosslinking tests were performed with Polymer 8. The pure polymer was crosslinked with a rate of 100 and 75 % - Polymer 8c1 and 8c2.

### 4.2.2  Pore generation

**Polymer 3**

A porous matrix of Polymer 3 - Polymer 3p1 - was produced as a prove of principle, for degradation studies and for X-ray computed tomography. The CT measurements performed showed a highly interconnected porous network similar to the matrices of Trimer 1p1 and 2p1. Thus, the method for pore generation can be used for trimers as well as for polymers. Table 4.1 and Figure 4.13 show the results of the CT measurement. Polymer 3p1 had the largest pores of the five measured samples. Consequently this polymer had the lowest number of pores, also showing a suitable matrix.

**Polymer 5**

Promising porous pellets could be produced from Polymer 5, giving Polymer 5p1. Polymer 5p1 was characterized by solid-state NMR as shown in Figure 4.14. The $^{31}$P NMR spectrum showed one broad peak as expected and in the $^{13}$C NMR spectroscopy one can clearly see that no peak for a C=C double bond in the range of 115-140 ppm is present. Moreover, SEM and

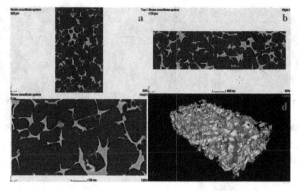

**Figure 4.13** – CT of Polymer **3p1**. 2D View from top, scale bar equals 400 μm (a), from right side, scale bar equals 250 μm (b) and from front side, scale bar equals 150 μm (c). 3D View of the block (d)

CT measurements were performed with Polymer **5p1** as shown in Figure 4.15 and Figure 4.16. The SEM shows a highly interconnected porous matrix. Interestingly, this pure polymer matrix seems to consist of small beads connected with each other. The pore size is mid-table of those tested. This polymer was also selected for biological testing.

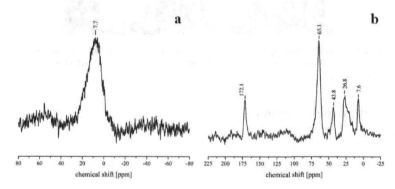

**Figure 4.14** – Solid-state NMR of Polymer **5p1**. $^{31}$P CP-MAS NMR shows a broad peak around 7.7 (a) and $^{13}$C CP-MAS NMR spectrum confirms the disappearance of the C=C (<5 %) signals in the range of 115-140 ppm (b). (172.1 ppm corresponds to C=O and 65.1 ppm to OCH$_2$ of the ester group, 43.8 ppm corresponds to NH−CH$_2$ of the glycine unit, 26.8 ppm comes from the thioether unit and 7.6 from CH$_3$ from trithiol)

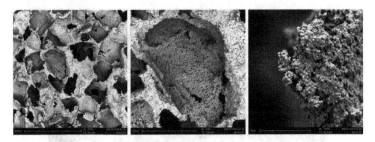

**Figure 4.15** – SEM of Polymer **5p1**. Scale bar equals $470\,\mu m$ (a), scale bar equals $120\,\mu m$(b) and scale bar equals $20\,\mu m$ (c)

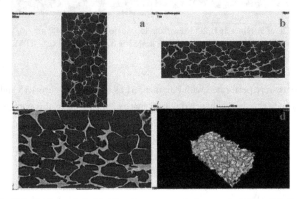

**Figure 4.16** – CT of Polymer **5p1**. 2D View from top, scale bar equals $400\,\mu m$ (a), from right side, scale bar equals $250\,\mu m$ (b) and from front side, scale bar equals $150\,\mu m$ (c). 3D View of the block (d)

### 4.2.3 Degradation studies

**Polymer 3c1**

Mass loss experiments were performed with Polymer **3c1** over a time of 27 weeks at two pH values, namely pH 2 for enhanced conditions and pH 7.4 for physiological conditions. This test was performed to compare the degradation of the allylamine trimer and polymer. In general, the degradation of Polymer **3c1** was very slow as expected. Polymer **3c1** kept in pH 2 showed a small mass loss of about 10 %, whereas pH 7.4 showed none as can be seen in Figure 4.17. This degradation rate is similar to the rate of Trimer **1c1** and thus, too slow. Besides, allylamine is toxic. Hence, this network cannot be used for tissue engineering.

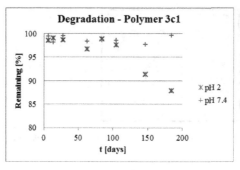

**Figure 4.17** – Degradation of Polymer **3c1** determined gravimetrically and performed at two different pH values, namely, 2 (HCl in H$_2$O) and 7.4 (Tris-buffer)

**Polymer 3p1**

The results of this experiment are shown in Chapter 4.3.3 to compare them with the corresponding blends.

# 4.3 Blends

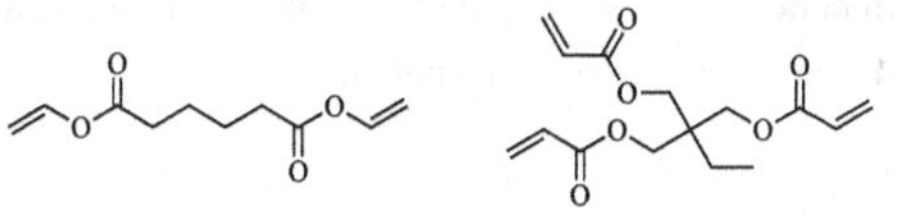

Divinyl adipate

Trimethylolpropane triacrylate

**Figure 4.18** – Blending agents

Divinyl adipate and trimethylolpropane triacrylate (herein after referred to as triacrylate) were used as blending agents. They were added in different ratios to the crosslinking reaction mixtures as a second double bond donor. Thereby, the chemical, biological and mechanical properties of the pellets were changed. This is of special interest for the porous matrices, since by changing the ratio of the blending agents in future the properties could be tailored for each cell type. Besides, the addition of blending agents has the advantage that less of the expensive and labourious produced poly(organophosphazene) is needed. Figure 4.19 shows a simplified schematic of a crosslinked polymer with divinyl adipate.

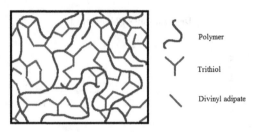

**Figure 4.19** – Simplified scheme of the network formed by crosslinking a polymer with divinyl adipate and trithiol

## 4.3.1   Crosslinking

**Polymer 5**

Polymer **5** was crosslinked 100 % in combination with 80 % divinyl adipate giving a nicely shaped pellet Polymer **5c2**. Since this reaction showed good results, porous material with this combination was produced.

**Polymer 8**

Polymer **8** was crosslinked in combination with 89 % divinyl adipate - Polymer **8c3** and 89 % triacrylate - Polymer **8c4**. They were quite soft, whereby Polymer **8c3** gave a nice shaped pellet, but Polymer **8c4** was too soft to obtain a pellet.

**Test substances**

These crosslinked networks were fabricated to be able to compare the properties of the pure and blended polymers with the pure blending agents. Therefore, pure triacrylate and divinyl adipate pellets were made, named Test substance **1** and **2** respectively.

## 4.3.2   Pore generation

**Trimer 1**

Porous pellets were fabricated using Trimer **1** in combination with 50 % triacrylate - Trimer **1p2**. After drying, the pellet collapsed in the middle. This reaction was performed as a proof of principle, but the products were not used for biological tests, as triacrylate may be harmful for the cells [108].

**Trimer 2**

Porous material was produced from a combination of Trimer 2 and 50 % triacrylate - Trimer **2p2**. The pellet collapsed in the middle after drying. Again, this reaction was performed as a prove of principle.

**Polymer 3**

Porous material of Polymer 3 in combination with triacrylate and divinyl adipate (Polymer **3p2** and **3p3**) was produced. The degradation rate of these materials was compared with Polymer **3p1** to find out, whether there is a difference in degradation between pure polymer material and blends and additionally, to compare porous polymer material with non-porous material (Polymer **3p1** versus Polymer **3c1**).

**Polymer 5**

Polymer 5 has a very promising structure, hence it was extensively used to prepare porous material. Materials with different polymer:divinyl adipate ratios were produced, namely 4:1, 1:1 and 1:4 giving Polymer **5p2**, **5p3** and **5p4** respectively. SEM and CT measurements of Polymer **5p4** were performed as shown in Figure 4.20 and Figure 4.21. A highly interconnected porous matrix similar to the pure polymer matrix in Figure 4.15 can be observed. In comparison to the pure polymer matrix, the surface of Polymer **5p4** is smooth and shows no bead-like structure. In Figure 4.21 showing the CT of Polymer **5p4** the cell walls seem thicker than in Figure 4.16 showing Polymer **5p1**. This is very likely not because of the addition of divinyl adipate, but rather because of the position, where the sample was cut out for measurement. On the bottom of the pellets the salt densified more, leaving less space for the polymer, whereas on the top there was more space between the salt crystals giving thicker walls. Biological tests were performed with Polymer **5p3** and **5p4** to determine, whether divinyl adipate in the mixture has an influence on cells.

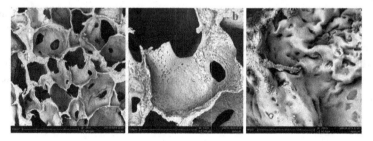

**Figure 4.20** – SEM of Polymer **5p4**. Scale bar equals 470 $\mu$m (a), scale bar equals 130 $\mu$m (b) and scale bar equals 40 $\mu$m (c)

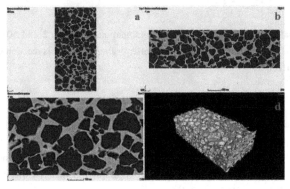

**Figure 4.21** – CT of Polymer **5p4**. 2D View from top, scale bar equals 400 $\mu$m (a), from right side, scale bar equals 250 $\mu$m (b) and from front side, scale bar equals 150 $\mu$m (c). 3D View of the block (d)

### Polymer 8

A combination of Polymer **8** and divinyl adipate (89 % of double bonds) was used to prepare porous scaffolds. This Polymer **8p1** had a suitable shape and was selected for biological testing.

### Test substances

Porous pellets of divinyl adipate were made, named Test substance **3**. Degradation studies and biological tests were performed with this test substance to have a comparison and a control.

## 4.3.3 Degradation studies

### Polymer 3p1, 3p2 and 3p3

Mass loss experiments were also performed with Polymer **3p1**, **3p2** and **3p3** over a time of 16 weeks at two pH values namely pH 2 for enhanced conditions and pH 7.4 for physiological conditions. This test was performed to compare the degradation of the porous allylamine polymer with its blends. No mass loss was observed in any of the samples within 16 weeks as shown in Figure 4.22. Thus, porous hydrophobic materials do not degrade much faster than non-porous hydrophobic materials. These polymer networks are not suitable for tissue engineering.

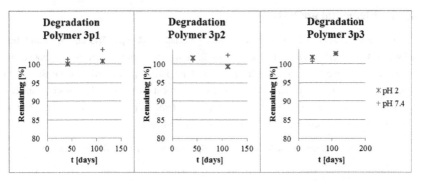

Figure 4.22 – Degradation Polymer 3p1, 3p2 and 3p3

**Polymer 5p1 and 5p3 and Test substance 3**

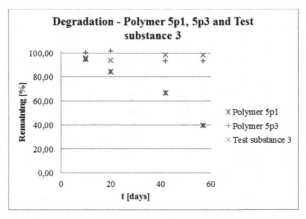

Figure 4.23 – Degradation Polymer 5p1 and 5p3 and Test substance 3

Mass loss experiments in deinoized H$_2$O (pH 6.2) were performed with Polymer **5p1** and **5p3** and Test substance **3** over a time of 57 days. Hence, the test series was comprised of a network with pure polymer with allyl glycinate as a side chain, with a 1:1 mixture of the polymer with divinyl adipate and pure divinyl adipate. Figure 4.23 shows that the network consisting of pure polymer was the fastest degrading of these three followed by the mixture. The pure divinyl adipate network shows no degradation. The blend degrades very slowly, so that only about 7 % degraded, whereas approximately 2/3 of the pure polymer network degraded in 57 days. The pure polymer network **5p1** provides an attractive degradation rate for cell and extracellular matrix growth [109], which can be slowed down by altering the polymer to divinyl adipate ratio. Hence, the degradation rate can be tailored to an optimum for each cell type.

#### 4.3.4 Elemental analysis

Elemental analysis of Polymers with allyl glycinate (**5p1, 5p2, 5p3** and **5p4**) and Test substance **3** was performed by J. Theiner. In Table 4.2 the results of the measurement, the theorectical values calculated with ChemDraw and the percentage of deviation are shown. In general, the measured values fit quite well to the theoretical ones. From these results, one can conclude, that the reactions worked as expected, the networks are quite homogeneous and that PEG-200 was completely removed by the washing procedure.

Table 4.2 – Elemental analysis of Polymer **5p1, 5p2, 5p3, 5p4** and Test substance **3**[a]

| Element | Polymer 5p1 | Polymer 5p2 | Polymer 5p3 | Polymer 5p4 | Test substance 3 |
|---|---|---|---|---|---|
| C found | 43.97 | 45.60 | 47.50 | 50.11 | 51.51 |
| C calc | 46.28 | 47.17 | 48.65 | 50.32 | 51.56 |
| C % deviation | 4.99 | 3.33 | 2.36 | 0.43 | 0.11 |
| H found | 6.21 | 6.31 | 6.54 | 6.85 | 6.86 |
| H calc | 7.30 | 7.28 | 7.26 | 7.23 | 7.21 |
| H % deviation | 14.93 | 13.39 | 9.99 | 5.33 | 4.92 |
| N found | 7.08 | 5.86 | 4.17 | 1.68 | - |
| N calc | 7.36 | 6.11 | 4.05 | 1.73 | - |
| N % deviation | 3.80 | 4.17 | 2.84 | 2.89 | - |
| S found | 11.23 | 11.46 | 12.36 | 11.97 | 12.75 |
| S calc | 11.23 | 11.66 | 12.37 | 13.17 | 13.76 |
| S % deviation | 0 | 1.76 | 0.08 | 9.15 | 7.34 |
| P found | 5.47 | 4.58 | 3.25 | 1.29 | - |
| P calc | 5.42 | 4.51 | 2.99 | 1.27 | - |
| P % deviation | 0.92 | 1.55 | 8.58 | 1.18 | - |

[a]X found is an average of two to three measurement results

## 4.4 Functionalization

All functionalizations performed gave good results showing that a single poly(organophosphazene) can easily be diversified by using thiol-ene photochemistry. Interestingly, crosslinking was hindered sterically neither by small nor by large functional groups added to the trimers and polymers. It was also very surprising that acidic and basic functionalities as for instance in glutathione did not show any hindrance in thiol-ene chemistry. The great advantage of this functionalization is that a small set of polymers or trimers can be produced, which can then easily be modified to tailor the mechanical, physical, chemical and biological porperties. Thus in the end, the properties may be ideally adjusted for each tissue type in tissue engineering.

## 4.4.1 Thioglycolic acid

Thioglycolic acid was the first functional group added by thiol-ene functionalization. It was used to test, whether a molecule can be added to a poly(organophosphazene) and whether it disturbs the crosslinking reaction afterwards. According to NMR and FTIR spectroscopy all thiol-ene reactions with thioglycolic acid were successful, thus further functionalization experiments were performed.

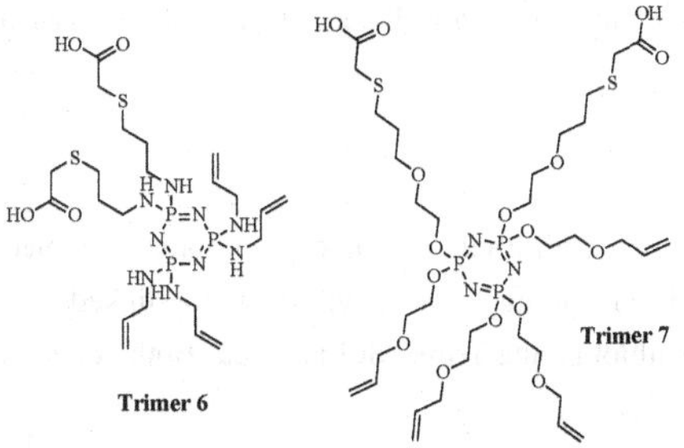

**Figure 4.24** – Trimer 1 and 2 modified with thioglycolic acid: Trimer 6 and Trimer 7, only one possible isomer is shown

### Trimer 6

The modification of Trimer 1 with thioglycolic acid lead to Trimer 6 as shown in Figure 4.24, which was further crosslinked with 100 and 75 % of the double bonds used by trithiol giving Trimer 6c1 and 6c2 as nicely shaped pellets.

### Trimer 7

Also the modification of Trimer 2 with thioglycolic acid giving Trimer 7, as shown in Figure 4.24, was successful. Trimer 7 could further be crosslinked with 75 % of the double bonds used by trithiol giving Trimer 7c1 as a rather soft, gel like pellet.

## 4.4.2 Ethyl-3-mercaptopropionate

The next functionalization agent used was ethyl-3-mercaptopropionate, as thioglycolic acid is toxic and therefore Trimer 6 or 7 could release toxic degradation products. As it was possible to add these rather small molecules, namely thioglycolic acid and ethyl-3-mercaptopropionate, it was decided to proceed with the biomolecule glutathione.

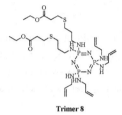

**Trimer 8**

**Figure 4.25** – Trimer 1 modified with ethyl-3-mercaptopropionate: Trimer 8, only one possible isomer is shown

**Trimer 8**

Trimer 1 was modified with 2 ethyl-3-mercaptopropionate molecules giving Trimer 8 as shown in Figure 4.25. The resulting Trimer 8 was then crosslinked with 100 % and 75 % of the double bonds used by trithiol giving Trimer **8c1** and **8c2**. Both pellets were very brittle.

### 4.4.3   Glutathione

Glutathione is a naturally occuring tripeptide used to balance the redox houshold of cells. It possesses cystein in the middle, hence, it is suitable for thiol-ene chemistry. The addition makes poly(organophosphazenes) more hydrophilic and should also lead to more attractive scaffolds for cells. Glutathione was also used as a pretest to optimize the conditions for the addition of the more expensive cyclic RGD, which is proven to increase adhesion of cells. Glutathione was successfully added to the poly(organophosphazenes) and with that, we showed that it is possible to functionalize with large molecules possessing different functional groups. Naturally, the properties of the final materials differ tremendously from the starting material, but still we were able to crosslink the functionalized material.

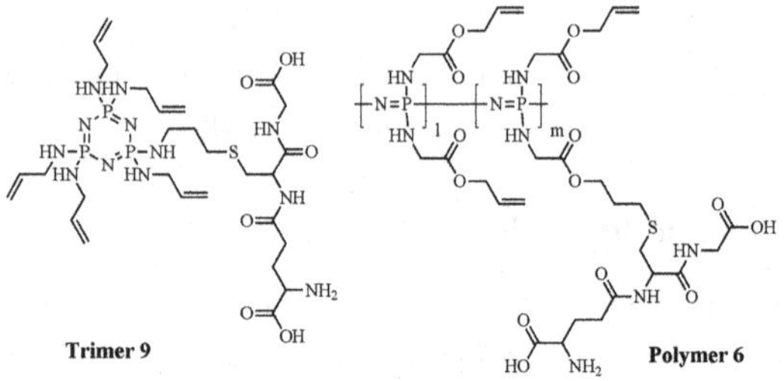

**Trimer 9**                **Polymer 6**

**Figure 4.26** – Trimer 1 and Polymer 5 modified with glutathione: Trimer 9 and Polymer 6

**Trimer 9**

Trimer **1** was modified with glutathione to give Trimer **9** as shown in Figure 4.26, with approximately 25 % of the double bonds substituted by the tripeptide. It is worth mentioning that by the addition of glutathione, Trimer **9** became hydrophilic enough to be soluble in water, what is not the case for Trimer **1**. Figure 4.27 shows the COSY-NMR spectrum, which clearly allows to distinguish between side chains with and without glutathione attached. It was also possible to crosslink Trimer **9** 100 % in a $H_2O$:THF (1:1) mixture giving Trimer **9c1** as a nice pellet. This is a quite interesting finding as several UV-photoreactions tried before were obviously disturbed by protic solvents like methanol and ethanol.

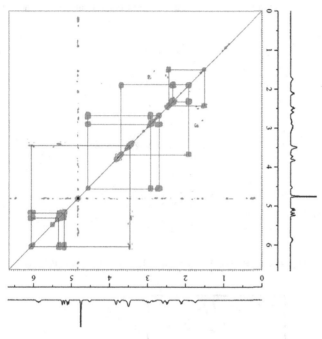

**Figure 4.27** – COSY-NMR measurement of Trimer **9**. The side chains with glutathione can clearly be distinguished from the side chains without glutathione: Peaks connected with red lines (square top right) are stemming from side chains reacted with glutathione, peaks connected with green lines (squares middle) are stemming from glutathione and peaks connected with blue lines (squares bottom left) are stemming from unreacted side chains.

**Polymer 6**

Polymer **6** was obtained by functionalization of Polymer **5** with approximately 25 % glutathione, giving Polymer **6** as shown in Figure 4.26. By the addition of glutathione to Polymer **5** the solubility was changed like in Trimer 9, but Polymer 6 was not completely soluble in H$_2$O, as it was not totally soluble in THF anymore. Nevertheless it was possible to crosslink Polymer 6 giving Polymer 6c1 as a pellet. This proved that the successful crosslinking of Trimer 9 in H$_2$O and THF was not an exception, but works with other poly(organophosphazenes) as well.

### 4.4.4 Fluorescein

Fluorescein isothiocyanate isomer I was modified to attain a thiol-linker to allow a thiol-ene functionalization. It was added in a small amount to Polymer 5 to visualize the distribution of the polymer in the 3D network. Both the pure polymer matrix and the one mixed with divinyl adipate showed a homogeneous distribution of poly(organophospazene)-chains in the porous networks.

**Polymer 7**

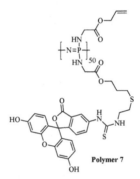

**Figure 4.28** – Polymer 5 modified with fluorescein derivative: Polymer 7

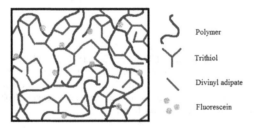

**Figure 4.29** – Simplified scheme of the network formed by crosslinking Polymer 7 functionalized with fluorescein with divinyl adipate and trithiol

Polymer 5 was functionalized with the modified fluorescein to give Polymer 7 as shown in Figure 4.28, which was immediately further reacted with trithiol to give a porous pellet namely

58

Polymer **7p1**. Furthermore, a pellet with divinyl adipate (1:1) was fabricated - Polymer **7p2**. Figure 4.29 represents a schematic drawing of this Polymer **7p2**. Impressively, significant parts of fluorescein survived the reaction in the UV-chamber and were still fluorescent afterwards. On Figure 4.30 the comparison of Polymer **5p3**, **7p2** and **7p1** (from left to right) under an UV-lamp (254 nm) can be seen.

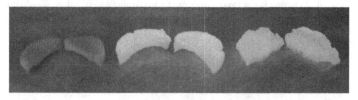

**Figure 4.30** – Cut pellets of Polymer **5p3**, **7p2** and **7p1** (from left to right) under an UV-lamp (254 nm). The fluorescein derivative is nicely distributed throughout the whole pellet in Polymer **7p2** and **7p1**

# 4.5 Biomedical tests

Here I summarized and explained an extract of the biological test results performed by our project partners with the trimer/polymer networks we produced.

## 4.5.1 Cytotoxicity studies

### Cell morphology studies

**Basic principle:**

The morphology of a cell is defined by the cell's shape, structure, form and size. ASC have a fibroblast-like shape and are attached to a substrate. Granularity around the nucleus and detachment are signs of deterioration, which could be caused by contamination, senescence, etc., but also by toxic substances in the culture medium [112]. Therefore, the morphology of cells can give an interesting insight in the health status of the cell culture.

### Results Red Cross - Trimer 1c1, 1c2, 1c3, 2c1, 2c2 und 2c3 as thin films

Conditioned media were obtained by washing the thin trimer films with culture medium for 72 hours (Medium 1) and after washing 4 times with PBS for 24 hours each, again washing with culture medium for 72 hours (Medium 2). Thus, if the trimers release any toxic substance, this can be found in the media. Then Medium 1 and 2 were added to healthy ASC cultures and the morphology was observed under a light microscope as shown in Figure 4.31 and Figure 4.32. Medium 1 of Trimer **1c2** and **1c3** had a toxic effect on the cells hindering cell growth, whereas

Medium 2 showed almost no effects. In Trimer **1c1** the effect of Medium 1 was less pronounced in Medium 2, but still obvious as can be seen in Figure 4.31. Medium 2 of Trimer **1c3** was contaminated and is therefore not shown here. So, some toxic compounds were washed out of the allylamine Trimer layers. Thus, careful washing of the polymers is essential. Both media from Trimer **2c1**, **2c2** and **2c3** were non-toxic and a normal cell morphology was observed. Ethylene glycol allyl ether is not toxic compared to allylamine and therefore should not lead to toxic degradation products. Hence, ethylene glycol allyl ether as a side chain is a better choice than allylamine.

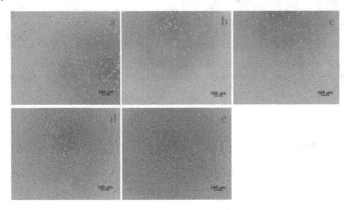

**Figure 4.31** – Light microscope images of Trimer **1c1**, **1c2** and **1c3**. (a) - (c) shows the cells cultured with Medium 1 and (d)-(e) with Medium 2, whereby (a) and (d) are samples from Trimer **1c1**, (b) and (e) from Trimer **1c2** and (c) from Trimer **1c3**

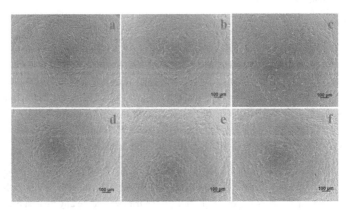

**Figure 4.32** – Light microscope images of Trimer **2c1**, **2c2** and **2c3**.(a) - (c) shows the cells cultured with Medium 1 and (d)-(f) with Medium 2, whereby (a) and (d) are samples from Trimer **2c1**, (b) and (e) from Trimer **2c2** and (c) and (f) from Trimer **2c3**

60

**LDH assay**

**Basic principle:**

Lactate dehydrogenase (LDH) is an enzyme usually present in the cytoplasm. When the plasma membrane is damaged, the cytoplasm and with it LDH is released to the culture medium. Therefore, LDH release is used as an indicator, whether for instance, a compound cultured with the cells is toxic. For the assay, cultivation medium and LDH reaction mixture were mixed. LDH released catalyzes the conversion of lactate to pyruvate via reduction of NAD+ to NADH. Then, NADH is used by diaphorase to reduce a tetrazolium salt (e.g. INT – 2-(4-iodophenyl)-3-(4-nitrophenyl)-5-phenyl-2H-tetrazolium chloride) to a red formazan product. The level of formazan can be determined spectrophotometrically and is directly proportional to the amount of LDH released.

**Results Red Cross - Trimer 1c1, 1c2, 1c3, 2c1, 2c2 and 2c3 as thin films**

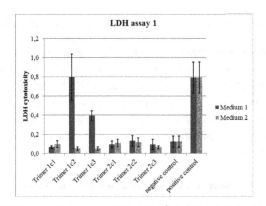

**Figure 4.33** – LDH cytotoxicity assay 1 of Trimer 1c1, 1c2, 1c3, 2c1, 2c2 and 2c3 as thin films

Again, conditioned Medium 1 and 2 as already explained in cell morphology studies were used. These media were applied to healthy ASC cultures. The media from the cell cultures were then mixed with the LDH reaction mixture (Pierce LDH Cytotoxicity Assay Kit) and gave the results shown in Figure 4.33. Media 1 of Trimer 1c2 and 1c3 were toxic, all others did not show any cytotoxicity. Thus, Trimers 1c2 and 1c3 could be freed from toxic substances by washing with medium and PBS. So again, washing is an important step to obtain tissue engineering material, which does not release toxic substances probably stemming from the chemical reactions. Trimer 1c1, 2c1, 2c2 and 2c3 did not release toxic substances within ten days.

**Results BioMed-zet - Trimer 1c1, 1c2, 1c3, 2c1, 2c2 and 2c3 as thin films**

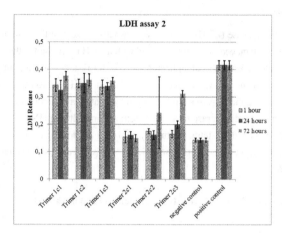

**Figure 4.34** – LDH cytotoxicity assay 2 of Trimer **1c1, 1c2, 1c3, 2c1, 2c2** and **2c3** as thin films

After sterilization of the trimer films with UV-light, they were washed three times with PBS. Then culture medium was conditioned on the plates for 1, 24 and 72 hours. These media mixed 1:1 with fresh media were given to cultivated HPMC. The following LDH assay gave the results shown in Figure 4.34. Trimer **1c1, 1c2** and **1c3** were toxic at every time point. Trimer **2c2** and **2c3** were toxic only at 72 hours and Trimer **2c1** was not toxic at all. From these results, one could conclude, that allylamine trimers are not suitable for this tissue engineering project and that a higher degree of crosslinking could be favourable.

**Results BioMed-zet - Trimer 1c1, 1c2 and 1c3 as powdered materials**

As the thin films of Trimer **1c1, 1c2** and **1c3** appeared to be toxic as shown in Figure 4.34, further investigations with powdered materials were carried out. Therefore, the material was washed for six days with culture medium, which was then applied on healthy HPMC. The LDH assay performed using media from day 1, 2 and 6 showed that Trimer **1c1, 1c2** and **1c3** are not toxic as shown in Figure 4.35. This proves that the materials itself were not toxic, but some leftovers from the process, which could not be washed out of the films.

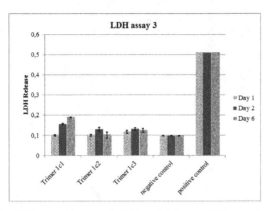

**Figure 4.35** – LDH cytotoxicity assay 3 of Trimer **1c1**, **1c2** and **1c3** as powdered materials

**Results BioMed-zet - Trimer 2p1 as porous pellet**

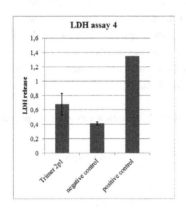

**Figure 4.36** – LDH cytotoxicity assay 4 of Trimer **2p1** as a porous pellet

The next step was to test HPMC directly on a scaffold. Therefore, 800 000 cells were seeded on Trimer **2p1**, and, as a control 80 000 cells were seeded in empty wells. The LDH assay performed after 24 hours showed that the level of LDH released was quite low, although ten times more cells were seeded on the scaffold than on the empty well. Thus, Trimer **2p1** is not toxic to cells on short terms (here 24 h). Figure 4.36 shows the results of this test.

**Results Red Cross - Polymer 5p1, 5p4 and 8p1 and Test substance 3 as porous pellets**

Prior to giving the samples to the biologists, they were washed for two days with a Soxhlet extraction with ethanol. The pellets were fragmented by the biologists and were washed twice

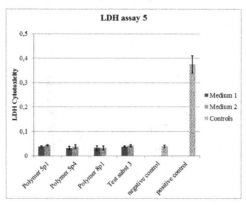

**Figure 4.37** – LDH cytotoxicity assay 5 of Polymer **5p1**, **5p4** and **8p1** and Test substance **3** as porous pellets

for 24 hours with culture medium resulting in Medium 1 and 2. These media were added to healthy ASC cultures. Then, the media from the cell cultures were mixed with the LDH reaction mixture. The results of the spectroscopic measurement are shown in Figure 4.37. None of the tested samples released toxic substances into the media. Hence, washing with Soxhlet extraction with ethanol helps to avoid toxic wash outs and the polymers themselves do not release toxic substances on short trials (here 24 hours).

**Results Red Cross - Polymer 5p1 degradation products of porous material**

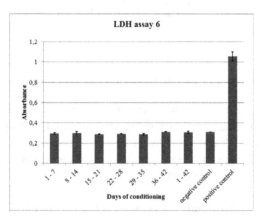

**Figure 4.38** – LDH cytotoxicity assay 6 of Polymer **5p1** conditioned medium after different time intervals

A porous pellet of Polymer **5p1** was cut into pieces and each was transferred in a small tube. Cell culture medium (1 ml) was added to each tube and they were incubated at 37 °C for 42 days. The medium of one tube was not changed in the whole period, whereas in the another tube the medium was changed every week. Several controls were performed as well. These media were applied to ASC and a LDH assay was performed as shown in Figure 4.38. None of the tested media was toxic, proving that Polymer **5p1**, which was the fastest degrading in our tests, does not release any toxic substances during degradation within 42 days.

## 4.5.2   Cell adhesion and proliferation studies

### Luminescent Cell Viability Assay

**Basic principle:**

ATP is the energy storage in all living organisms, hence presents the metabolic activity of cells. The number of viable cells in a culture can be determined based on quantitation of ATP present in cells. Therefore, the CellTiter-Glo©Reagent is added to the cultured cells. Thereby, the cells are lysed and ATP is released to the medium, where it is used up by the enzyme luciferase. Luciferase converts beetle luciferin into oxyluciferin, whereby a "glow-type" luminescence is generated. The amount of luminescene produced is proportional to the amount of ATP in the medium and the number of cells present in culture is directly proportional to the amount of ATP measured. [113]

**Results Red Cross - Polymer 5p1, 5p3 and 5p4 as a porous pellet**

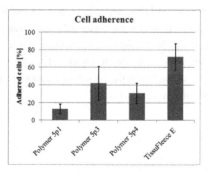

**Figure 4.39** – Cell adhesion of ASC on Polymer **5p1**, **5p3** and **5p4** using a CellTiter-Glo©-Luminescent Cell Viability Assay (Promega, WI)

The porous pellets of Polymer **5p1**, **5p3** and **5p4** were cut in pieces of 4 mm³ in size and 500 000 cells were seeded on each piece. TissuFleece E (Baxter, Heidelberg, Germany) was used as a control [109]. It is a collagen type I sponge from bovine or equine origin, which

65

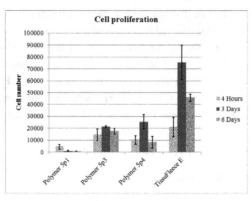

**Figure 4.40** – Proliferation studies of ASC on Polymer **5p1**, **5p3** and **5p4** using a CellTiter-Glo©- Luminescent Cell Viability Assay (Promega, WI)

possesses pores in a similar size range (188±76 $\mu$m) [114] as our polymers. It recently out-performed several clinically used biomaterials in terms of cell proliferation and metabolic ac-itivity. [115], and hence, is a reasonable control scaffold. The number of ASC adhered on the matrices was measured 4 hours and ASC proliferation was determined 3 and 6 days after cell seeding using the CellTiter-Glo©assay. The luminescence was measured with an Infinite 200 microplate reader (TECAN, Maenndorf, Switzerland). Three replicates were used for the analysis and measurements were conducted in duplicates. The results in Figure 4.39 show that 12.6±5.9 % of the cells adhere to Polymer **5p1**, whereas 41.8±19.0 % and 30.6±11.6 % adhere to the divinyl adipate blends Polymer **5p3** and **5p4** respectively. The number of metabolic active ASC decreased on Polymer **5p1**, but increased 1.5 and 2.5-fold within the first 3 days on Polymer **5p3** and **5p4** respectively as shown in Figure 4.40. A decrease of ASC after 6 days was observed in all four samples, probably due to contact inhibition, because of high cell density. This results show that these polymers are promising candidates for tissue engineering, but future work, in particular surface functionalizations will be necessary to increase cell adhesion. Further studies are under current investigation.

# 5.  Conclusion and future work

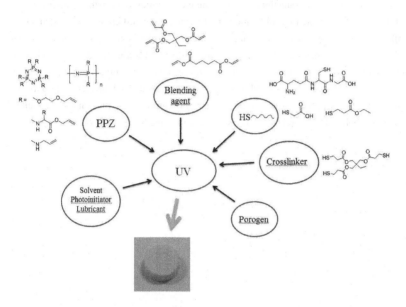

**Figure 5.1** – Scheme of the modular hybrid system established. Only a small set of poly(organo-phosphazenes) (PPZ) combined with blending and functionalization agents creates a tunable system, able to adapt to different microenvironments necessary for certain types of tissue.

This project demonstrates that poly(organophosphazenes) are promising candidates for tissue engineering matrices. It was shown that biocompatible polymers with reasonable degradation rates could be produced. Thiol-ene photochemistry can be used to crosslink the material to gain mechanical strength necessary to carry the cell load. Additionally, thiol-ene photochemistry can be used to add functional moieties, with which not only the chemical properties of the matrix can be changed, but also cell-recognition signals can be added to increase cell adhesion and tailor cell migration, differentiation and proliferation. A photocrosslinking particulate-leaching method was developed to obtain a highly interconnected porous structure to enable cell invasion and communication. The degradation rate and the mechanical properties can further be tailored by the addition of blending agents in different ratios. Thus, a modular hybrid system was established as shown in the schematic drawing Figure 5.1.

The next steps towards a tissue engineering scaffold include the incorporation of bioactive factors. The addition of glutathione proved that this is possible. Often short peptide sequences, which are known to enhance cell adhesion are used. The most prominent candidate of such an amino acid sequence is RGD. Especially cyclic RGD molecules have been found to increase cell adhesion tremendously [98]. A version of a cyclic RGD possessing a cystein exists, which could be added via photofunctionalization. The incorporation of growth factors, hormones or cytokines prone to increase proliferation and lineage-specific differentiation of stem cells [6], will be a more demanding task to be fulfilled in future. Furthermore, the salt leaching technique is limited to small matrices. For larger organ sized scaffolds a more advanced technique, as for example a solid freeform fabrication, which uses a layer-manufacturing strategy to create a scaffold from computer-generated models, must be applied. With this techniques a vascular system can be mimicked, allowing a natural supply of nutrients and oxygen and the removal of waste metabolites [5], what is a crucial point for up-scaling the tissue to organ size.

# Bibliography

[1] C. T. Laurencin, A. M. Ambrosio, M. D. Borden, and J. A. Cooper. Tissue engineering: orthopedic applications. *Annual review of biomedical engineering*, 1:19–46, 1999.

[2] R. Langer and J. Vacanti. Tissue engineering. *Science*, 260(5110):920–926, 1993.

[3] L. G. Griffith and G. Naughton. Tissue engineering–current challenges and expanding opportunities. *Science (New York, N.Y.)*, 295(5557):1009–1014, 2002.

[4] T. Dvir, B. P. Timko, D. S. Kohane, and R. Langer. Nanotechnological strategies for engineering complex tissues. *Nature Nanotechnology*, 6(1):13–22, 2010.

[5] E. Sachlos and J. T. Czernuszka. Making tissue engineering scaffolds work. Review on the application of solid freeform fabrication technology to the production of tissue engineering scaffolds. *European Cells and Materials*, (5):29–40, 2003.

[6] H. Naderi, M. M. Matin, and A. R. Bahrami. Review paper: Critical Issues in Tissue Engineering: Biomaterials, Cell Sources, Angiogenesis, and Drug Delivery Systems. *Journal of Biomaterials Applications*, 26(4):383–417, 2011.

[7] C. W. Patrick, A. G. Mikos, and L. V. McIntire. *Frontiers in tissue engineering: Prospectus of tissue engineering*. Pergamon, Oxford and U.K and New York and NY and U.S.A, 1st ed edition, 1998.

[8] J. Shi, A. R. Votruba, O. C. Farokhzad, and R. Langer. Nanotechnology in Drug Delivery and Tissue Engineering: From Discovery to Applications. *Nano Letters*, 10(9):3223–3230, 2010.

[9] L. E. Freed and G. Vunjak-Novakovic. Culture of organized cell communities. *Advanced Drug Delivery Reviews*, 33(1-2):15–30, 1998.

[10] W. H. Eaglstein and V. Falanga. Tissue engineering and the development of Apligraf, a human skin equivalent. *Clinical Therapeutics*, 19(5):894–905, 1997.

[11] J. Vacanti. Tissue engineering and regenerative medicine: from first principles to state of the art. *Journal of pediatric surgery*, 45(2):291–294, 2010.

[12] R. E. Horch. Future perspectives in tissue engineering: ?Tissue Engineering? Review Series. *Journal of Cellular and Molecular Medicine*, 10(1):4–6, 2006.

[13] R. Langer. Tissue Engineering. *Molecular Therapy*, 1(1):12–15, 2000.

[14] A. Khademhosseini, R. Langer, J. Borenstein, and J. P. Vacanti. Microscale technologies for tissue engineering and biology. *Proceedings of the National Academy of Sciences*, 103(8):2480–2487, 2006.

[15] K. Y. Tsang, M. C. H. Cheung, D. Chan, and K. S. E. Cheah. The developmental roles of the extracellular matrix: beyond structure to regulation. *Cell and Tissue Research*, 339(1):93–110, 2010.

[16] B. M. Lewin. *Genes IX.* Jones and Bartlett Publ., Sudbury and Mass. [u.a.], [9. ed.] edition, 2008 [erschienen 2007.

[17] R. O. Hynes. The extracellular matrix: not just pretty fibrils. *Science (New York, N.Y.),* 326(5957):1216–1219, 2009.

[18] R. Sasisekharan, Z. Shriver, G. Venkataraman, and U. Narayanasami. Roles of heparan-sulphate glycosaminoglycans in cancer. *Nature reviews. Cancer,* 2(7):521–528, 2002.

[19] D. W. Hutmacher. Scaffold design and fabrication technologies for engineering tissues — state of the art and future perspectives. *Journal of Biomaterials Science, Polymer Edition,* 12(1):107–124, 2001.

[20] I. Armentano, M. Dottori, E. Fortunati, S. Mattioli, and J. M. Kenny. Biodegradable polymer matrix nanocomposites for tissue engineering: A review. *Polymer Degradation and Stability,* 95(11):2126–2146, 2010.

[21] A. Arem. Collagen Modifications. *Clinics in Plastic Surgery,* 12(2):209–220, 1985.

[22] C. Shi, Y. Zhu, X. Ran, M. Wang, Y. Su, and T. Cheng. Therapeutic Potential of Chitosan and Its Derivatives in Regenerative Medicine. *Journal of Surgical Research,* 133(2):185–192, 2006.

[23] M. Vert, J. Mauduit, and S. Li. Biodegradation of PLA/GA polymers: increasing complexity. *Biomaterials,* 15(15):1209–1213, 1994.

[24] Y. Wang, G. A. Ameer, B. J. Sheppard, and R. Langer. A tough biodegradable elastomer. *Nature biotechnology,* 20(6):602–606, 2002.

[25] A. L. Bauer, T. L. Jackson, Y. Jiang, and A. Czirók. Topography of Extracellular Matrix Mediates Vascular Morphogenesis and Migration Speeds in Angiogenesis. *PLoS Computational Biology,* 5(7):e1000445, 2009.

[26] D. E. Discher. Tissue Cells Feel and Respond to the Stiffness of Their Substrate. *Science,* 310(5751):1139–1143, 2005.

[27] K. R. Levental, H. Yu, L. Kass, J. N. Lakins, Mi. Egeblad, J. T. Erler, S. F. T. Fong, K. Csiszar, A. Giaccia, W. Weninger, M. Yamauchi, D. L. Gasser, and V. M. Weaver. Matrix crosslinking forces tumor progression by enhancing integrin signaling. *Cell,* 139(5):891–906, 2009.

[28] N. D. Evans, C. Minelli, E. Gentleman, V. LaPointe, S. N. Patankar, M. Kallivretaki, X. Chen, C. J. Roberts, and M. M. Stevens. Substrate stiffness affects early differentiation events in embryonic stem cells. *European cells & materials,* 18:1–13; discussion 13–4, 2009.

[29] T. Rozario and D. W. DeSimone. The extracellular matrix in development and morphogenesis: a dynamic view. *Developmental biology,* 341(1):126–140, 2010.

[30] E. D. Cohen, K. Ihida-Stansbury, M. M. Lu, R. A. Panettieri, P. L. Jones, and E. E. Morrisey. Wnt signaling regulates smooth muscle precursor development in the mouse lung via a tenascin C/PDGFR pathway. *The Journal of clinical investigation,* 119(9):2538–2549, 2009.

[31] J.-S. Silvestre, B. I. Lévy, and A. Tedgui. Mechanisms of angiogenesis and remodelling of the microvasculature. *Cardiovascular research,* 78(2):201–202, 2008.

[32] R. R. Chen, E. A. Silva, W. W. Yuen, and D. J. Mooney. Spatio-temporal VEGF and PDGF delivery patterns blood vessel formation and maturation. *Pharmaceutical research,* 24(2):258–264, 2007.

[33] D. C. Darland and P. A. D'Amore. Blood vessel maturation: vascular development comes of age. *The Journal of clinical investigation,* 103(2):157–158, 1999.

[34] T. P. Richardson, M. C. Peters, A. B. Ennett, and D. J. Mooney. Polymeric system for dual growth factor delivery. *Nature biotechnology*, 19(11):1029–1034, 2001.

[35] J. A. Jansen, J. W. M. Vehof, P. Q. Ruhé, H. Kroeze-Deutman, Y. Kuboki, H. Takita, E. L. Hedberg, and A. G. Mikos. Growth factor-loaded scaffolds for bone engineering. *Journal of controlled release : official journal of the Controlled Release Society*, 101(1-3):127–136, 2005.

[36] S. Koch, C. Yao, G. Grieb, P. Prével, E. M. Noah, and G. C. M. Steffens. Enhancing angiogenesis in collagen matrices by covalent incorporation of VEGF. *Journal of materials science. Materials in medicine*, 17(8):735–741, 2006.

[37] G. C. M. Steffens, C. Yao, P. Prével, M. Markowicz, P. Schenck, E. M. Noah, and N. Pallua. Modulation of angiogenic potential of collagen matrices by covalent incorporation of heparin and loading with vascular endothelial growth factor. *Tissue engineering*, 10(9-10):1502–1509, 2004.

[38] S. A. DeLong, J. J. Moon, and J. L. West. Covalently immobilized gradients of bFGF on hydrogel scaffolds for directed cell migration. *Biomaterials*, 26(16):3227–3234, 2005.

[39] A. H. Zisch, S. M. Zeisberger, M. Ehrbar, V. Djonov, C. C. Weber, A. Ziemiecki, E. B. Pasquale, and J. A. Hubbell. Engineered fibrin matrices for functional display of cell membrane-bound growth factor-like activities: study of angiogenic signaling by ephrin-B2. *Biomaterials*, 25(16):3245–3257, 2004.

[40] J. J. Moon, S.-H. Lee, and J. L. West. Synthetic biomimetic hydrogels incorporated with ephrin-A1 for therapeutic angiogenesis. *Biomacromolecules*, 8(1):42–49, 2007.

[41] H. R. Allcock. Inorganic-Organic Polymers. *Advanced Materials*, 6(2):106–115, 1994.

[42] H. R. Allcock. The synthesis of functional Polyphosphazenes and their Surfaces. *Applied Organometallic Chemistry*, 12:659–666, 1998.

[43] H. R. Allcock and K. B. Visscher. Preparation and characterization of poly(organophosphazene) blends. *Chemistry of Materials*, 4(6):1182–1187, 1992.

[44] H. R. Allcock, K. B. Visscher, and I. Manners. Polyphosphazene-organic polymer interpenetrating polymer networks. *Chemistry of Materials*, 4(6):1188–1192, 1992.

[45] H. R. Allcock. *Chemistry and applications of polyphosphazenes*. Wiley-Interscience, Hoboken and N.J, 2003.

[46] H. Henke, S. Wilfert, A. Iturmendi, O. Brüggemann, and I. Teasdale. Branched polyphosphazenes with controlled dimensions. *Journal of Polymer Science Part A: Polymer Chemistry*, 51(20):4467–4473, 2013.

[47] S. Wilfert, H. Henke, W. Schoefberger, O. Brüggemann, and I. Teasdale. Chain-End-Functionalized Polyphosphazenes via a One-Pot Phosphine-Mediated Living Polymerization. *Macromolecular rapid communications*, page 10.1002/marc.201400114, 2014.

[48] H. R. Allcock. A Perspective of Polyphosphazene Research. *Journal of Inorganic and Organometallic Polymers and Materials*, 16(4):277–294, 2007.

[49] Hornbaker, E.D. and Li, H.M. Process for preparing low molecular weight linear phosphonitrilic chloride oligomers, patent: Us4198381 a, 1980.

[50] G. D'Halluin, R. de Jaeger, J. P. Chambrette, and P. Potin. Synthesis of poly(dichlorophosphazenes) from Cl3P:NP(O)Cl2. 1. Kinetics and reaction mechanism. *Macromolecules*, 25(4):1254–1258, 1992.

[51] D. P. Tate. Polyphosphazene elastomers. *Journal of Polymer Science: Polymer Symposia*, 48(1):33–45, 1974.

[52] M. Deng, S. G. Kumbar, Y. Wan, U. S. Toti, H. R. Allcock, and C. T. Laurencin. Polyphosphazene polymers for tissue engineering: an analysis of material synthesis, characterization and applications. *Soft Matter*, 6(14):3119, 2010.

[53] C. T. Laurencin, C. D. Morris, H. Pierre-Jacques, E. R. Schwartz, A. R. Keaton, and L. Zou. Osteoblast culture on bioerodible polymers: studies of initial cell adhesion and spread. *Polymers for Advanced Technologies*, 3(6):359–364, 1992.

[54] C. T. Laurencin, M. E. Norman, H. M. Elgendy, S. F. El-Amin, H. R. Allcock, S. R. Pucher, and A. A. Ambrosio. Use of polyphosphazenes for skeletal tissue regeneration. *Journal of biomedical materials research*, 27(7):963–973, 1993.

[55] S. Lakshmi, D.S Katti, and C.T Laurencin. Biodegradable polyphosphazenes for drug delivery applications. *Advanced Drug Delivery Reviews*, 55(4):467–482, 2003.

[56] J. Crommen, J. Vandorpe, and E. Schacht. Degradable polyphosphazenes for biomedical applications. *Journal of Controlled Release*, 24(1-3):167–180, 1993.

[57] S. Sethuraman, L. S. Nair, S. El-Amin, R. Farrar, M.-T. N. Nguyen, A. Singh, H. R. Allcock, Y. E. Greish, P. W. Brown, and C. T. Laurencin. In vivo biodegradability and biocompatibility evaluation of novel alanine ester based polyphosphazenes in a rat model. *Journal of biomedical materials research. Part A*, 77(4):679–687, 2006.

[58] H. Allcock. Synthesis and characterization of pH-sensitive poly(organophosphazene) hydrogels. *Biomaterials*, 17(23):2295–2302, 1996.

[59] M. J. Kade, D. J. Burke, and C. J. Hawker. The power of thiol-ene chemistry. *Journal of Polymer Science Part A: Polymer Chemistry*, 48(4):743–750, 2010.

[60] C. E. Hoyle, T. Y. Lee, and T. Roper. Thiol-enes: Chemistry of the past with promise for the future. *Journal of Polymer Science Part A: Polymer Chemistry*, 42(21):5301–5338, 2004.

[61] J. A. Codelli, J. M. Baskin, N. J. Agard, and C. R. Bertozzi. Second-generation difluorinated cyclooctynes for copper-free click chemistry. *Journal of the American Chemical Society*, 130(34):11486–11493, 2008.

[62] K. L. Killops, L. M. Campos, and C. J. Hawker. Robust, Efficient, and Orthogonal Synthesis of Dendrimers via Thiol-ene "Click" Chemistry. *Journal of the American Chemical Society*, 130(15):5062–5064, 2008.

[63] S. Caldwell, D. W. Johnson, M. P. Didsbury, B. A. Murray, J. J. Wu, S. A. Przyborski, and N. R. Cameron. Degradable emulsion-templated scaffolds for tissue engineering from thiol–ene photopolymerisation. *Soft Matter*, 8(40):10344, 2012.

[64] A. B. Lowe. Thiol-ene "click" reactions and recent applications in polymer and materials synthesis. *Polymer Chemistry*, 1(1):17, 2010.

[65] H. R. Allcock, M. V. B. Phelps, E. W. Barrett, M. V. Pishko, and W.-G. Koh. Ultraviolet Photolithographic Development of Polyphosphazene Hydrogel Microstructures for Potential Use in Microarray Biosensors. *Chemistry of Materials*, 18(3):609–613, 2006.

[66] Y.-C. Qian, X.-J. Huang, C. Chen, N. Ren, X. Huang, and Z.-K. Xu. A versatile approach to the synthesis of polyphosphazene derivatives via the thiol-ene reaction. *Journal of Polymer Science Part A: Polymer Chemistry*, 50(24):5170–5176, 2012.

[67] H. R. Allcock and A. G. Scopelianos. Synthesis of sugar-substituted cyclic and polymeric phosphazenes and their oxidation, reduction, and acetylation reactions. *Macromolecules*, 16(5):715–719, 1983.

[68] N.R. Krogman, A. L. Weikel, N. Q. Nguyen, L. S. Nair, C. T. Laurencin, and H. R. Allcock. Synthesis and Characterization of New Biomedical Polymers: Serine- and Threonine-Containing Polyphosphazenes and Poly(1-lactic acid) Grafted Copolymers. *Macromolecules*, 41(21):7824–7828, 2008.

[69] N. L. Morozowich, A. L. Weikel, J. L. Nichol, C. Chen, L. S. Nair, C. T. Laurencin, and H. R. Allcock. Polyphosphazenes Containing Vitamin Substituents: Synthesis, Characterization, and Hydrolytic Sensitivity. *Macromolecules*, 44(6):1355–1364, 2011.

[70] A. K. Salem, R. Stevens, R. G. Pearson, M. C. Davies, S. J. B. Tendler, C. J. Roberts, P. M. Williams, and K. M. Shakesheff. Interactions of 3T3 fibroblasts and endothelial cells with defined pore features. *Journal of biomedical materials research*, 61(2):212–217, 2002.

[71] J. Guan, J. J. Stankus, and W. R. Wagner. Biodegradable elastomeric scaffolds with basic fibroblast growth factor release. *Journal of controlled release : official journal of the Controlled Release Society*, 120(1-2):70–78, 2007.

[72] G. Wei, Q. Jin, W. V. Giannobile, and P. X. Ma. Nano-fibrous scaffold for controlled delivery of recombinant human PDGF-BB. *Journal of controlled release : official journal of the Controlled Release Society*, 112(1):103–110, 2006.

[73] E. A. Botchwey, M. A. Dupree, S. R. Pollack, E. M. Levine, and C. T. Laurencin. Tissue engineered bone: measurement of nutrient transport in three-dimensional matrices. *Journal of biomedical materials research. Part A*, 67(1):357–367, 2003.

[74] D. Brancazio, J.F. Bredt, M. Cima, A. Curodeau, T. Fan, and E. Sachs. Three dimensional printing system, patent: Us5807437 a, 1998.

[75] C. W. Hull. Method for production of three-dimensional objects by stereolithography, patent: Us4929402 a, 1990.

[76] S. Scott Crump. Apparatus and method for creating three-dimensional objects, patent: Us5121329 a, 1992.

[77] R. Landers and R. Mülhaupt. Desktop manufacturing of complex objects, prototypes and biomedical scaffolds by means of computer-assisted design combined with computer-guided 3D plotting of polymers and reactive oligomers. *Macromolecular Materials and Engineering*, 282(1):17–21, 2000.

[78] J. L. Forsyth, K. F. Philbrook, and R. C. Sanders Jr. 3-D model maker, patent: Us5506607 a, 1996.

[79] J. D. Hartgerink, E. Beniash, and S. I. Stupp. Self-assembly and mineralization of peptide-amphiphile nanofibers. *Science (New York, N.Y.)*, 294(5547):1684–1688, 2001.

[80] S. Zhang. Fabrication of novel biomaterials through molecular self-assembly. *Nature biotechnology*, 21(10):1171–1178, 2003.

[81] D. J. Mooney, D. F. Baldwin, N. P. Suh, J. P. Vacanti, and R. Langer. Novel approach to fabricate porous sponges of poly(d,l-lactic-co-glycolic acid) without the use of organic solvents. *Biomaterials*, 17(14):1417–1422, 1996.

[82] H. Lo, M. S. Ponticiello, and K. W. Leong. Fabrication of controlled release biodegradable foams by phase separation. *Tissue engineering*, 1(1):15–28, 1995.

[83] I. V. Yannas, J. F. Burke, P. L. Gordon, C. Huang, and R. H. Rubenstein. Design of an artificial skin. II. Control of chemical composition. *Journal of biomedical materials research*, 14(2):107–132, 1980.

[84] C. J. Doillon, C. F. Whyne, S. Brandwein, and F. H. Silver. Collagen-based wound dressings: control of the pore structure and morphology. *Journal of biomedical materials research*, 20(8):1219–1228, 1986.

[85] K. Whang, C. H. Thomas, K. E. Healy, and G. Nuber. A novel method to fabricate bioabsorbable scaffolds. *Polymer*, 36(4):837–842, 1995.

[86] M. Borden, M. Attawia, Y. Khan, and C. T. Laurencin. Tissue engineered microsphere-based matrices for bone repair. *Biomaterials*, 23(2):551–559, 2002.

[87] J. L. Brown, L. S. Nair, and C. T. Laurencin. Solvent/non-solvent sintering: a novel route to create porous microsphere scaffolds for tissue regeneration. *Journal of biomedical materials research. Part B, Applied biomaterials*, 86(2):396–406, 2008.

[88] R. C. Thomson, M. J. Yaszemski, J. M. Powers, and A. G. Mikos. Fabrication of biodegradable polymer scaffolds to engineer trabecular bone. *Journal of Biomaterials Science, Polymer Edition*, 7(1):23–38, 1996.

[89] C. T. Laurencin, S. F. El-Amin, S. E. Ibim, D. A. Willoughby, M. Attawia, H. R. Allcock, and A. A. Ambrosio. A highly porous 3-dimensional polyphosphazene polymer matrix for skeletal tissue regeneration. *Journal of Biomedical Materials Research*, 30:133–138, 1996.

[90] A. G. Mikos, G. Sarakinos, S. M. Leite, J. P. Vacant, and R. Langer. Laminated three-dimensional biodegradable foams for use in tissue engineering. *Biomaterials*, 14(5):323–330, 1993.

[91] J. Jagur-Grodzinski. Polymers for tissue engineering, medical devices, and regenerative medicine. Concise general review of recent studies. *Polymers for Advanced Technologies*, 17(6):395–418, 2006.

[92] G. A. Silva, C. Czeisler, K. L. Niece, E. Beniash, D. A. Harrington, J. A. Kessler, and S. I. Stupp. Selective differentiation of neural progenitor cells by high-epitope density nanofibers. *Science (New York, N.Y.)*, 303(5662):1352–1355, 2004.

[93] H.-W. Jun and J. L. West. Modification of polyurethaneurea with PEG and YIGSR peptide to enhance endothelialization without platelet adhesion. *Journal of biomedical materials research. Part B, Applied biomaterials*, 72(1):131–139, 2005.

[94] S. P. Massia and J. A. Hubbell. Vascular endothelial cell adhesion and spreading promoted by the peptide REDV of the IIICS region of plasma fibronectin is mediated by integrin alpha 4 beta 1. *The Journal of biological chemistry*, 267(20):14019–14026, 1992.

[95] K. von der Mark and J. Park. Engineering biocompatible implant surfaces. *Progress in Materials Science*, 58(3):327–381, 2013.

[96] H. Shin, S. Jo, and A. G. Mikos. Biomimetic materials for tissue engineering. *Biomaterials*, 24(24):4353–4364, 2003.

[97] D. J. Leahy, I. Aukhil, and H. P. Erickson. 2.0 å Crystal Structure of a Four-Domain Segment of Human Fibronectin Encompassing the RGD Loop and Synergy Region. *Cell*, 84(1):155–164, 1996.

[98] J. Zhu, C. Tang, K. Kottke-Marchant, and R. E. Marchant. Design and Synthesis of Biomimetic Hydrogel Scaffolds with Controlled Organization of Cyclic RGD Peptides. *Bioconjugate Chemistry*, 20(2):333–339, 2009.

[99] Y. Luo and M. S. Shoichet. A photolabile hydrogel for guided three-dimensional cell growth and migration. *Nature materials*, 3(4):249–253, 2004.

[100] K.-B. Lee, D. J. Kim, Z.-W. Lee, S. I. Woo, and I. S. Choi. Pattern Generation of Biological Ligands on a Biodegradable Poly(glycolic acid) Film. *Langmuir*, 20(7):2531–2535, 2004.

[101] I. Freeman and S. Cohen. The influence of the sequential delivery of angiogenic factors from affinity-binding alginate scaffolds on vascularization. *Biomaterials*, 30(11):2122–2131, 2009.

[102] M. Prambauer. Synthesis and Characterization of Polyphosphazenes for Tissue Engineering Applications; Diploma thesis, 2013.

[103] N. L. Morozowich, R. J. Mondschein, and H. R. Allcock. Comparison of the Synthesis and Bioerodible Properties of N-linked versus O-linked Amino Acid Substituted Polyphosphazenes: Manuscript draft.

[104] C. Rim and D. Y. Son. Facile and efficient synthesis of star-shaped oligomers from a triazine core. *Tetrahedron Letters*, 50:4161–4163, 2009.

[105] B. Wang. Development of a One-Pot in Situ Synthesis of Poly(dichlorophosphazene) from PCl 3. *Macromolecules*, 38(2):643–645, 2005.

[106] H. R. Allcock, C. A. Crane, C. T. Morrissey, J. M. Nelson, S. D. Reeves, C. H. Honeyman, and I. Manners. Living Cationic Polymerization of Phosphoranimines as an Ambient Temperature Route to Polyphosphazenes with Controlled Molecular Weights. *Macromolecules*, 29:7740–7747, 1996.

[107] T. Kaufmann, M. T. Gokmen, S. Rinnen, H. F. Arlinghaus, F. Du Prez, and B. J. Ravoo. Bifunctional Janus beads made by "sandwich" microcontact printing using click chemistry. *Journal of Materials Chemistry*, 22(13):6190, 2012.

[108] J. Autian. Structure-Toxicity Relationships of Acrylic Monomers. *Environmental Health Perspectives*, 11:141–152, 1975.

[109] F. Hildner, S. Concaro, A. Peterbauer, S. Wolbank, M. Danzer, A. Lindahl, P. Gatenholm, H. Redl, and M. van Griensven. Human Adipose-Derived Stem Cells Contribute to Chondrogenesis in Coculture with Human Articular Chondrocytes. *Tissue Engineering Part A*, 15(12):3961–3969, 2009.

[110] I. Teasdale, S. Wilfert, I. Nischang, and O. Brüggemann. Multifunctional and biodegradable polyphosphazenes for use as macromolecular anti-cancer drug carriers. *Polymer Chemistry*, 2(4):828, 2011.

[111] S. Wilfert, A. Iturmendi, W. Schoefberger, K. Kryeziu, P. Heffeter, W. Berger, O. Brüggemann, and I. Teasdale. Water-soluble, biocompatible polyphosphazenes with controllable and pH-promoted degradation behavior. *Journal of Polymer Science Part A: Polymer Chemistry*, 52(2):287–294, 2014.

[112] invitrogen and gibco. *Cell Culture Basics: Handbook: http://www.vanderbilt.edu/viibre/CellCultureBasics-EU.pdf.* 26.02.2014.

[113] Promega Corporation. CellTiter-Glo Luminescent Cell Viability Assay Technical Bulletin, TB288. 16.04.2014.

[114] O. J. Buettner. *TissuFleece E als Trägermatrix im Knochen Tissue Engineering: Evaluation of the collagen sponge TissuFleece E as carrier for osteoblasts in bone tissue enginnering.* PhD thesis, Albert-Ludwigs-Universität Freiburg, Breisgau, 2006.

[115] M. Gierloff, T. Nitsche, S. Adam-Klages, K. Liebs, J. Hedderich, V. Gassling, J. Wiltfang, D. Kabelitz, and Y. Asil. In vitro comparison of different carrier materials with rat bone marrow MSCs. *Clinical oral investigations*, 18(1):247–259, 2014.